The Sixth Wave

How to succeed in a resource-limited world

James Bradfield Moody
Bianca Nogrady

16pt

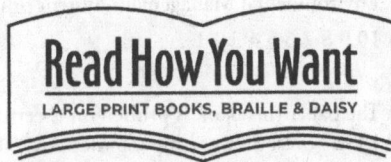

Copyright Page from the Original Book

A Vintage book
Published by Random House Australia Pty Ltd
Level 3, 100 Pacific Highway, North Sydney NSW 2060
www.randomhouse.com.au
First published by Vintage in 2010

Addresses for companies within the Random House Group can be found at www.randomhouse.com.au/offices

National Library of Australia
Cataloguing-in-Publication Entry
Moody, James Bradfield.
The sixth wave / James Bradfield Moody, Bianca Nogrady.
ISBN 9781741668896 (pbk.)
Human capital – Forecasting.
Natural resources – Forecasting.
Technological innovations.
Other Authors/Contributors: Nogrady, Bianca.
303.49

Cover image by iStockphoto
Cover design by Christabella Designs
Typeset in Adobe Garamond Pro 13.5/16 pt
Printed in Australia by Griffin Press, an Accredited ISO AS/NZS 14001:2004 Environmental Management System printer.

10 9 8 7 6 5 4 3 2 1

TABLE OF CONTENTS

For Jasper and Nina

Note

Except when stated otherwise, all quotations in this book come from interviews conducted by the authors.

Introduction

Prediction is a dangerous game. Not only is it difficult in foresight and easy in hindsight, but it also carries the possibility that you'll be completely and utterly wrong. This book makes a very bold prediction indeed. It predicts the next wave of innovation.

The act of predicting the future does have some advantages, whether you're right or not. Predictions give you something to test your ideas against and can put some of the big and little things you see every day into a broader context. Our prediction will do just this.

In this book we'll think about big things, such as rising energy and food prices and the debate about climate change, and the way these factors are shaping investments in clean technologies, now accounting for the most venture-capital investment of any sector in the United States. And smaller things, such as the race between Google and Microsoft to take millions of photos of some of the largest cities in the world and create 3D images from these photographs. This would allow you to visit New York, enter a department store, and browse and buy its goods – all from your home computer.

And then there are the very little things that happen on a big scale. Try this small experiment. If you have a phone equipped with the technology called Bluetooth, go to a crowded place and use your phone to search for other Bluetooth devices. You will find a large number of other phones, each emitting its identification number. Now, imagine that it wasn't just telephones you could see, but every product and every item around you.

What do all these things have in common? They are all part of the same trend – a trend that will fundamentally transform human society: The sixth wave of innovation.

The first half of this book will look at the five previous waves of innovation, and how they are shaped by market forces, technologies and the very glue that holds societies together. We will see how we can learn from these to identify the sixth great wave and how it will create more opportunity than we can possibly imagine.

For a prediction to be useful, you need to know how to act on it. The second half of this book outlines what we call 'sixth-wave thinking', and explains how investors, businesses, communities and governments can create more opportunity, wealth and competitive advantage by embracing this way of thinking. One of the keys to creating value and succeeding in the sixth wave will be to better understand our natural, social and financial resources.

This book is for anyone who is interested in thinking about what a sustainable future holds for humankind, and in understanding how the massive changes currently taking place in the world fit together. In particular, it is for those who wish to understand and take advantage of the shifts that are occurring around us.

In making our bold prediction, we will take you on a journey through the next thirty years of global development, and we will reach into all areas of human endeavour. The sixth wave is

on its way and the world will never be
the same again.

Part I

The next wave of innovation

Part 1

Introduction

Two hundred and fifty years ago, this book would have been written by candlelight on paper made of pulped rags using a quill dipped in ink. The authors themselves would have been a lot less healthy, less mobile, less educated and most likely would have been living and working in cramped and unsanitary conditions. The changes that have swept our world since those times are too numerous to even begin to document, but if we step back from the minutiae, we can see a pattern emerge from the apparent chaos of reform.

Since the Industrial Revolution, the tide of progress has ebbed and flowed in cycles known as 'long waves' of innovation. These waves transform society, economies and industry almost beyond recognition, in half-century cycles of disruption and opportunity, followed by saturation and downturn.

The first half of our journey through the next wave of innovation starts by exploring the very nature of innovation itself – what it is and how it works. We will think about the three pillars of innovation, and how the things that are most often thought of as the 'drivers' of innovation are only part of the story.

We then look at the five long waves of innovation that have shaped the world over the last two hundred years. From milling and steam power to steel and electricity, and from mass production to information and communication technologies, the innovation that defined these waves not only shaped the technology of each era, but also had a profound impact on social structures, resource consumption and world trade.

To understand these waves of innovation, we will look at the one that we are most familiar with – the information and communication technology wave, which began in the 1970s. We'll see that it wasn't simply the new technology that drove this wave; and we will examine where the money that paid for the technology came from, and why companies such as Microsoft, eBay, Amazon and Google were inevitable from the time that the first microprocessors were made.

We'll also explore what it means to be on the cusp of a new wave, experiencing the transition from one wave to another. These cusps are associated with massive global depression – which is bad – but the disruption also heralds enormous opportunities.

So what is the sixth wave of innovation? Put simply, it is a revolution that will see our world transformed from one heavily addicted to the consumption of resources, to a world in which resource-efficiency is the name of the game. With the planet running out of natural resources and staring down the barrel of climate change and food security issues, the sixth wave will see humanity finally make the break away from resource-dependence. Everything, from the smallest tree and light switch to the largest cities and online communities, will have a measurable value and economic growth will no longer be tied to resource consumption or waste production.

In this next wave of innovation, resource scarcity and massive inefficiencies will be the big market opportunities. Waste will be the source of this opportunity and nature will be our source of inspiration and competitive advantage.

The way we organise the institutions which make up our society will also be transformed. The increasing competition for natural resources will pressure us to account for every tonne of carbon, joule of energy and litre of water. Things that until now have been valueless will acquire price tags, from carbon to water to biodiversity. The sixth wave of innovation will bring an explosion of tools that allow us to monitor, map and manage the resources around us.

And driving this all will be a spectacular boom in technologies ranging from clean technology to digital mapping to online collaboration. Traditional physical and geographical boundaries will mean nothing in a world where everything and everyone is online. Industry will increasingly realise value from services rather than resource-intensive products, and new business leaders will emerge to challenge the status quo.

On our journey into the next wave we will meet some inspiring individuals who are leading the charge in this new future. From electric cars to carbon farming, kite power to landfill mining, they are at the front of the wave. These pioneers have seen the opportunity ahead of them and are already beginning to act, showing us a little of what the future will hold.

But first things first: let's get back to basics...

Chapter 1

Why do things keep changing?

Change always happens both slower and faster than we expect. Slower, because the next breakthrough – whether it's the Dick Tracy phone-watch or a cure for cancer – always seems to be just around the corner, and faster, because when change does happen it comes in a great rush, and before you know it there's a whole new world out there.

Meet Allister. Allister is a thirty-something IT professional working for a large professional services firm in Brisbane, Australia. He manages the software for the company's finance systems and keeps its IT support running smoothly. He and his wife have a nice house in the suburbs and have just welcomed their first child, William, into the world.

Allister works in a high-technology job and lives in a high-technology environment, but his world hasn't really changed very much over the past few years. His job is much the same as the one he was doing five years ago, and many of the products in his house are made by companies he has grown up with. Most of the everyday technologies that he uses have been around for much of his life, such as his television, internal-combustion-engine car, microwave and computer. There's not that much difference between how he used his computer ten years ago and how he uses it today; his company even uses some of the same enterprise software as it did in 2002.

But take a step back from this picture and you begin to appreciate that there is actually quite a lot of change going on around Allister; in fact, it's fair to say his life has been transformed. When he was born, computers were behemoths that existed only in the basements of large organisations or high-tech laboratories – now they are in every household and on just about every desk. Technologies that his parents would have dismissed as science fiction – mobile phones, GPS receivers and iPods – have become reality. If Allister asked his father about the brand names that dominated the market before Allister was born, chances are he would only recognise a small number of them. Allister's standard of living is also considerably higher, by all objective measures, than that of his parents: his life expectancy is at least ten years longer than his father's,[1] he has spent more years in education and he has greater disposable income. Even more significantly, when Allister was born thirty-odd years ago, his current field of work didn't even exist.

What makes Allister and everyone in his age bracket so interesting is that they were born when things like computers and the internet were in their formative stages. Their births coincided with the start of a huge technological leap forward that has transformed daily life almost beyond recognition. Their generation has barely known, and probably couldn't even imagine, a world without computers, mobile phones and the internet.

But even more interesting is what is going to happen to Allister's baby boy, William. Like his father, William is going to see the world transformed. Allister and William have more than their genes in common – they were both born around the beginning of a wave of innovation.

What is this thing called innovation?

Put very simply, *innovation* is a new way of doing things. But it is about more than just a new gadget – it's about the entire, wonderful process of technological change. This includes not only the technology but also the creativity behind it, the entrepreneurship in getting it to market, and the shifts in thinking that encourage the market to embrace it.

We can talk about types of innovation in many different ways – by industry, by the degree of its novelty or by the depth of its impact. There can be product innovations, which change the things we use, or process innovations, where the change is in how products are made. Innovation can occur as relative change (when something better and/or cheaper comes along; for example, the printing press compared to hand transcription), or as absolute change (which delivers something entirely new; for example, the discovery of penicillin). Innovation can be spectacularly sudden and transformative, or it can creep up on us in increments.

Innovation can nurture and sustain the status quo by helping us to adapt to changing circumstances – for example, by improving the fuel efficiency of cars and thus enabling us to continue driving our cars in the same way even as oil prices rise – or it can totally disrupt the way we do things. It can be continuous or discontinuous in how it affects traditional ways of doing things. A continuous innovation provides a seamless transition from one state to the next, such as from one version of an internet browser to the next upgrade, whereas a discontinuous innovation requires us to abandon the previous system in favour of a new one. Innovation can also bring about modular changes – changes that affect one part of a system at a time – or broader architectural changes, which transform an entire system at once.[2]

There are some constants, however. When we look more closely at innovation, the changes that enable and accompany it fall into three main categories:

1) Development of new *technologies;*
2) Changes in the *market,* or the *demand for new or existing technologies;* and
3) Changes in the *institutions* that link, enable and encourage the first two components to come together.

The process of innovation occurs if you change just one of these three areas. Indeed, scientists, sociologists and innovation theorists have argued for decades about whether innovation is driven by *science-push* – simply put, 'if you build it, they will come' – or *demand-pull,* the idea that if enough people want something to be red, for instance, someone will come up with a way to make it red. But it is now generally agreed that the truth lies somewhere in between: that there is a flow of information and knowledge between and within companies, from science to marketing to the end-users.[3] Innovation is a systemic process, in which the linkages are at least as important as the actors.

But true and lasting innovation happens when all three – technologies, markets and institutions – change together. It's not enough simply to have a new technology or a new market demand; each factor is dependent on the other two to really make things happen. But whether it is the market, the technologies or the institutions that really drive the innovation can depend on the situation.

So what actually happens in each of these three key components that enables great leaps forward in innovation? The first category – changes in technology – is probably the easiest to understand. A perfect example is the mobile phone, which didn't exist in any recognisable form until the early 1970s. Now nearly half the world's population owns one.

The next key ingredient for innovation is the market – more specifically, changes in what consumers need or want. One example of this is the meteoric rise of the 'short message service', or SMS, on mobile phones. The SMS was originally dismissed as a relatively insignificant feature of the mobile phone network, and indeed many original carriers disabled the option, thinking it would be more trouble to maintain than it was worth. What these carriers underestimated was the appeal of texting rather than talking. We might enjoy having a good chat over the phone, but almost as useful, and fun, is the ability to send an 'asynchronous' message – a message that does not require the sending and receiving to happen at the same time. Suddenly, mobile phones became more than a device enabling two people to talk to each other while on the move. A new and highly lucrative market emerged.

While technology and markets are very important factors in driving innovation, the most significant development actually tends to occur *between* these two components. This is what we call *institutions,* which describes the linkages between markets and technologies – the systems, structures and regulations.

Institutions are tough to create and even tougher to measure and manage – many are very complicated. They have been loosely defined as the 'humanly devised constraints that structure political, economic and social interaction'.[4] Institutions are critical to most of our societies; for example, a society without the institution of property rights would find it hard to have bank loans.

Once again, the mobile phone industry can help illustrate the importance of institutions. Having new technology like a mobile phone and a market crying out for it is all well and good, but mobile phones couldn't really take off until governments allowed mobile phone towers to be built in cities. The network infrastructure for mobile phones – towers, transmitters and switches – had to be interfaced with the existing phone system and workable pricing plans had to be developed before mobile phones could make any progress in the market. More interestingly, whole patterns of behaviour had to emerge around mobile phone use, such as when it was appropriate to make a phone call and when it was not. All of these institutions had to be established before the mobile communications industry could be truly successful.

These institutions can make a big difference to the success or failure of a new technology in the market. Pricing structures for mobile phones in the United States were very different to those in Europe – in the US people had to pay not just to make a call from their mobile phone, but to receive one as well. This meant people were much more reluctant to give out their mobile phone number in case they found themselves paying to receive calls from people they didn't really want to hear from. This single factor put a huge brake on the spread of mobile phone use throughout the US.

So, while technology and markets are sexier concepts, and easier to understand, most innovation actually happens in institutions, where existing technologies meet existing markets in new ways. Economists look at markets, scientists and engineers look at technology, but innovation theorists look to the institutions that bring them together.

It's all about timeframes

Cast your mind back about one hundred years, to the early 1900s. Imagine you have suddenly come into some money, and you're looking to invest it. You have two choices: a company that sells horses, or a company making one of these newfangled automobile things. Which do you think is going to give you the biggest bang for your investment buck? With the benefit of hindsight, the answer seems obvious, but one hundred years ago this would have been a genuine dilemma.

Let's look at the two technologies in turn. If you lived in England or the United States in the early 1900s, you would be very familiar with the technology known as the horse and buggy. It was reliable, in that it would start every day without fail, and you would probably be pretty familiar with its user interface, having learnt how to control a horse from an early age. This technology also had other benefits: it ran on biofuel (hay, grass and the occasional lump of sugar) and was itself ultimately biodegradable. There was, admittedly, a small problem with waste, which caused some consternation in big cities, but in general the horse got you where you needed to go with a minimum of fuss.

The car, on the other hand, was a totally different matter. In the early 1900s it had a number of drawbacks. It was smelly and unreliable and was often difficult to start in cold weather. Its user interface was alien to most people, and it also emitted a serious amount of noise and air pollution. While it had the potential to go faster than a horse and buggy, there were road rules that prohibited it from travelling any faster than a horse walking at a moderate pace. And, most significantly, unlike the biofuelled horse, a car required the driver to buy petrol from inconvenient locations.

In short, you would have faced the choice of investing in either a reliable, predictable, familiar, clean, cheap and biodegradable mode of transport, or a new, uncertain, unreliable, dirty, gas-guzzling mode of transport that forced you to travel to places you didn't even want to go to in the first place to buy fuel.

Side by side, the usefulness of these two technologies in 1900 was comparable at best. But when viewed across a longer timeframe, the horse had little or no potential to evolve beyond the basic four-legs-and-a-saddle model (at least until you start considering it in evolutionary timeframes). But the car, as history has shown, was at the very beginning of its 'technology trajectory', and has almost completely transformed itself over the course of that time. Had you invested in four-legged transport you'd be kicking yourself, but if you'd taken a punt on the motorcar, chances are there would be a type of vehicle named after you by now.

The point is that changes in markets, technologies and institutions do not occur in a temporal vacuum. To be successful, innovation must take into account the past, present and future. If we compare two technologies without taking their trajectories into account, we miss a lot of the picture. What if you had to make the choice today between the internal-combustion engine and the electric vehicle? Which trajectory is each of these on?

When thinking about innovation, historical context and the potential for growth in the future are just as important as utility in the here and now.

Innovation across the ages

The Russian economist Nikolai Dimitrievich Kondratiev was born at the wrong time and in the wrong place. His life started out promisingly enough. Born in 1892 into a peasant family just north of Moscow, he was educated at the University of St Petersburg and worked his way up to the position of Minister of Supply in the Russian government at the tender age of twenty-five.

He went on to found and direct an influential research organisation, the Institute of Conjecture, and develop a five-year plan for Soviet agriculture. But in 1928 the prevailing political winds had changed and things started to turn sour for Kondratiev. He was bumped from his directorship of the Institute, accused by the precursor of the KGB of being a member of a non-existent political party, and soon the Soviet Premier, Joseph Stalin, was calling for his head.

Kondratiev kept his head but was forced to serve eight years in a gulag. Shortly before his sentence was up, he was tried again, as part of Stalin's Great Purges, and was sentenced to a further ten years of isolation. Unfortunately, someone somewhere missed the memo from that day's trials, and on the same day his sentence was handed down – 17 September 1938 – the forty-six-year-old Kondratiev was executed by firing squad.

As tragic as his story is, it's not the reason Kondratiev is well known among economists. His most significant claim to fame was his discovery that Western capitalist economies appear to undergo distinct cycles of boom and bust, each lasting around fifty or sixty years. The first of these cycles started in Britain in around 1780 – the Industrial Revolution.

Based on his study of fluctuations in interest rates and the prices of commodities, Kondratiev characterised a growth phase, in which prices increased and interest rates were low, followed by a deceleration phase, which saw a drop in prices and an increase in interest rates. He published his observations in his 1925 book, *The Major Economic Cycles.*

But Kondratiev didn't necessarily make the connection between these economic cycles and innovation. Another economist, Joseph Schumpeter, picked up where Kondratiev had left off and incorporated Kondratiev's cycles into his developing theory of innovation. In a nod to the original source, he called the cycles 'Kondratiev waves'. In his *Theory of Economic Development,* Schumpeter identified a number of what he called 'successive industrial revolutions' that tended to feature an initial period of turmoil and instability, followed by a period of growth as dominant technological designs emerged.[5]

These waves all experience a common evolution. During the initial phases of a wave, the technologies involved are quite disruptive. There is increasing conflict between traditional industry players and a whole spawn of new businesses. Entrepreneurs have a field day as they exploit opportunities, weaknesses and inadequacies in the business models of established companies.

Eventually, financiers start to take notice of the returns offered by these emerging businesses and begin investing heavily in the new technologies. Dominant designs emerge built around technological winners (for example, Microsoft Windows), and large companies that are nimble enough soon realise that significant changes to their business models are required (think of IBM's shift towards services and away from products). Throughout this period, those who had the foresight to invest in the new technologies make a killing.

But the party doesn't last forever. Eventually, the returns on investment begin to slow down. The potential of the shiny new technology is becoming fully realised, and the feisty young upstarts settle down to become the more staid establishment. This is known as a turning point or saturation point and has been coupled with massive downturns in the global economy – known variously as a crash, depression or crisis. Interestingly, in this stage of a long wave, history has shown that financiers begin to attempt more drastic financial manoeuvres to stem the decline in returns, trying to squeeze blood out of a stone.[6] But all this hardship isn't for nothing – it helps institutions evolve to fit the new technological paradigm and enables the opportunities of the next wave to begin to emerge.

And that is the basic anatomy of a Kondratiev wave – disruption and frenzy to saturation and maturation. So far, there have been five distinct Kondratiev waves over the past two centuries. The first wave coincides with the period of history now known as the Industrial Revolution. Mainly thanks to water-powered mechanisation, industry took a proverbial great leap forward with the help of new technologies. One was the 'spinning jenny', a multis-pool yarn-spinning machine that combined with a water-powered cotton mill; another was the metallurgical process called 'puddling', by which pig-iron could be mass-converted into steel. These technological advances both enabled and were encouraged by a shift away from home-based cottage industries and towards business partnerships, entrepreneurships and factory-style mass production.[7]

The second wave was powered by steam and is often described as the 'age of the railways'. In England, the Liverpool–Manchester railway opened in 1830 as the first intercity passenger train line powered entirely by steam engines, although it was mainly used to transport raw materials and finished products to and from the Liverpool docks and Manchester textile mills.

Kondratiev's third wave was the era of electricity, heavy engineering and steel. On 4 September 1882, Thomas Edison threw the switch at the Pearl Street Station in New York, thought to be the world's first electricity power station, and the lives of a few select

New Yorkers in Lower Manhattan suddenly got a little brighter. Alexander Graham Bell's invention of the telephone enabled communication over greater distances, while steel production shifted up a gear thanks to the introduction of the Bessemer process, which enabled the highly useful alloy to be mass-produced cheaply.

In the fourth wave, the motorcar emerged. Henry Ford's Highland Park Plant – the first assembly-line factory for automobiles – began production in 1913, but it wasn't until after the Great Depression that the motor car truly came into its own. The oil-based economy also took off, aided by a new process of converting crude oil into petrol that enabled oil companies to more than double their petrol output.[8] The fourth wave witnessed a move to mass-production on a scale never before seen.

The fifth and most recent Kondratiev wave – the information and communication technology wave – began in the 1970s with the emergence of silicon chip technology. It has seen an explosion in computing power and network capability and was given the catchy name of the information revolution.

Perhaps the most striking feature of these waves is that they didn't only result in a shift in the technologies used; they also led to quite profound societal changes. Each wave changed many of our fundamental institutions – from the way that businesses were organised to the way children were educated. It was these societal shifts, even more so than the changes in technology, that saw the waves often described as technological 'revolutions'.

Summary of Kondratiev waves

	Wave 1: Cotton, Iron and Water Power	Wave 2: Railways, Steam Power and Mechanisation	Wave 3: Steel, Heavy Engineering and Electrification	Wave 4: Oil, Automobiles and Mass Production	Wave 5: Information and Communication Technology
Upswing[1]	1780s–1815	1848–1873	1895–1918	1941–1973	1980–2001
Downswing[1]	1815–1848	1873–1895	1918–1940	1973–?	2001–?
Technologies[1]	Cotton-spinning and iron products, water-wheels, bleach	Railways and railway equipment, steam engines, machine tools, alkali industry	Electrical equipment, heavy engineering, heavy chemicals, steel products	Automobiles, trucks, tractors, tanks, diesel engines, aircraft, oil refineries	Computers, software, telecommunication equipment, biotechnology
Core inputs[1]	Iron, raw cotton, coal	Iron, coal	Steel, copper, metal alloys	Oil, gas, synthetic materials	Integrated circuits
Transport and communications infrastructure[1]	Canals, turnpike roads, sailing ships	Railways, telegraph, steam ships	Steel railways, steel ships, telephone	Radio, motorways, airports, airlines	Internet, 'Information highways'
Corporate organisation[2]	Owner-manager	Hierarchy	Division	Matrix	Network linkages

1. Freeman, C., Louca, F., As Time Goes By: From the Industrial Revolutions to the Information Revolution, Oxford University Press,

Oxford, 2002. | 2. Linstone, H. 'Corporate planning, forecasting and the long wave', Futures 34: 317–336, 2002

Boom and bust

Unfortunately, all good things must come to an end. Waves of innovation and technological progress eventually slow down, leading – as history thus far has shown – to economic depression.

These depressions happen during global turning points as the frenzy of investment in the wave slows down. For example, Kondratiev placed the turning point for the Industrial Revolution around 1815, the end of the Napoleonic wars. Following a boom in railroad construction the turning point for the second wave was the Long Depression, a period of general deflation and low growth following a panic in 1873, triggered by the collapse of Philadelphia banking firm Jay Cooke & Company. The waning of the third Kondratiev wave – a period when electricity, heavy engineering and steel made powerhouses of many economies – led to the proportionately spectacular crash of the Great Depression, while the fourth wave – mass production – reached saturation with the 1973 oil shocks and 1973–1974 stock market crash.

The dominant way of doing something can change significantly during these downturns, and for business it is very much a case of 'evolve or perish'. For example, General Electric is the only company still standing of the twelve that listed on the newly formed Dow Jones Industrial Average of the New York Stock Exchange in 1896. Another stayer is the computer giant IBM, but to achieve its unusual longevity it has had to reinvent itself numerous times to adjust to each new wave. It is currently in the process of doing so again.

Schumpeter described this as 'creative destruction'; as Chris Papenhausen puts it, this is 'the notion that new products and processes, in part, are based on the death of old products and processes'.[9] While these downswings result in economic depression, the situation is far from depressing; this period is actually a time of great opportunity. In between the waves of innovation, a great deal of change can occur. The old ways of doing things gradually become obsolete, institutions and structures that have stymied innovation by clinging to the status quo weaken, and methods once considered traditional and therefore sacrosanct are finally challenged.[10] The situation can become quite turbulent as the technological advances that built and sustained the previous wave start to grow stale and limited. But while a turbulent environment can cause some linkages to be redundant, it also creates an opportunity for new and exciting linkages to form.

These new linkages give rise to innovation. New combinations of concepts, resources and technologies can bring surprising outcomes. Sometimes these linkages form between quite unexpected entities, creating even more opportunities for innovation in directions that might not have been thought of before. For example, when you join a mobile phone together with a digital camera, you don't just get a phone that can also take pictures – you have created a device that can send photo messages around the globe.

In Kondratiev's and Schumpeter's original thinking, each wave lasted around fifty or sixty years. These long timeframes were due to factors such as the long life of investments in infrastructure and the time it took for economies and societies to evolve, known as 'institutional inertia'. However, things are now speeding up, and while the information and communication wave has only really had thirty years of life so far, all the indications are that we are in a period of transition. Could it be that we are experiencing the downswing of the fifth wave of innovation – and witnessing the birth of the sixth wave?

What causes these waves?

Kondratiev waves are not just random fluctuations; as we have said before, each wave is marked by a massive shift in the three key elements of markets, institutions and technology. These waves come about amidst the emergence of a cluster of new and rapidly emerging technologies, such as those that were built around the steam engine, motorcar or mobile phone. But technology is not enough; there must also be a need – a large and expanding market – for these technologies. Finally, there must be a way for these technologies to connect to and be adopted by the market; these are the institutions that emerge with these new technologies and markets.

Let's take a closer look at the second wave – the age of railways. The most significant new technology to emerge was the steam engine, which enabled people to travel far greater distances, in a fraction of the time and far more safely, than was possible with the steam engine's predecessor, the horse-drawn coach. This in turn created exciting new opportunities for trade with partners once considered too far away, and opened up new markets for new products. It also prompted the institutional uptake of better long-distance communications, which then paved the way for the rise of the telephone.

While the big-picture view of these Kondratiev waves suggests a smooth path of innovation, at the coalface it's a highly chaotic and disrupted environment. Entrepreneurs and innovators thrive in environments characterised by uncertainty and change. Creativity is heightened in times of turbulence. When regulations, technologies and society are in a state of flux, opportunities emerge for the creation of new products and new markets. At a time when old rules and practices are no longer applicable, we are forced to look for new ways of doing things.

A period of transition?

Which brings us to the present. As this book is being written, the world is experiencing what many commentators view as the worst global downturn since the Great Depression.

Some innovation theorists originally speculated that the turning point of the fifth wave was the dotcom bust of 2001.[11] While not as spectacular as some of the global depressions of the past, this had all of the hallmarks of the end of a frenzy of investment and spectacular returns in the information revolution. However, following this bust the developed and developing world experienced a period of unprecedented growth. This growth was fuelled by a combination of the maturing returns from the fruits of information and communication technologies and overheated investments in financial services, property prices and heightened consumerism. Some developing economies – such as those of China and India – were successful in bringing vast numbers of people out of poverty. But it was not to last.

The sixteenth of September, 2008, will be remembered as a sobering day for the world economy. The global banking giant Lehman Brothers announced that it was going to close its doors immediately, sending shockwaves throughout the global financial system. It was the biggest bankruptcy in corporate history – on that single day, more than US$500 billion of value was wiped off the combined London and New York stock exchanges. Fear that the global financial system was in meltdown started to spread.

Over the following year things went from bad to worse. In the United States, GDP contracted sharply by an annualised rate of 6.3 per cent in the December quarter. Every US state contracted in the month of February, the first time this had occurred since 1979. Industrial production in Japan decreased by 9.4 per cent in February 2009, finishing 37.6 per cent lower than a year earlier – its fifth consecutive fall. The International Monetary Fund predicted that global bank losses would hit US$4.1 trillion.

This book is not so much about the global financial crisis as what this crisis could herald. It is Schumpeter's 'creative destruction' on a scale that even he could not have envisaged. Could it be that, like other great depressions, the economic downturn of 2008 and 2009 heralds the transition from one wave of innovation to another?

The next wave

Allister was born blissfully unaware of the innovation storm raging above his baby head, as the fourth Kondratiev wave came to its end and the fifth emerged from its ashes. Allister's son William is, no doubt, sleeping peacefully despite a similar transition occurring during his babyhood. But with the timeframes of Kondratiev waves shrinking, Allister is in the unique position of being able to witness the transitions between two consecutive waves of innovation – fourth to fifth, and fifth to sixth.

This book is about the next wave, the one that will follow the information and communication technology wave. Before getting to that, however, let's turn to history and ask what really happened during the last wave of innovation. What might we learn as we witness the saturation point of the fifth wave and dawn of the sixth?

Chapter 2

The last wave

Readers of the 15 November 1971 edition of *Electronic News* probably had little inkling that they were witnessing history in the making. Buried inside the pages of this trade rag was an advertisement that heralded a revolution – one that was to transform our way of life completely.

The ad was for Intel's new programmable chip, the Intel 4004 microprocessor, which the company proudly hailed as 'a micro-programmable computer on a chip'. Two years earlier, the Nippon Calculating Machine Corporation had approached Intel to design twelve custom chips for its new printing calculator. The idea was that each of these chips would control a separate element of the calculator, but Intel engineers Marcian 'Ted' Hoff, Federico Faggin and Stan Mazor had other ideas. They came up with a set of just four chips, including a unique central processing unit, the 4004.

It was the first commercial chip that could be custom-programmed for use in a range of products – a 'chip of all trades'. Intel must have had some idea of its enormous potential because they lowered their fee for producing the chip in exchange for the rights to the design and to market it for other uses. Initially intended for use in something as small and simple as a calculator, this chip laid the foundation for the fifth wave of innovation – the information and communication technology wave.

The 'new economy'

Look around you. If you're at home, you may have a computer, a printer and a telephone on the desk, a television, a DVD player, a stereo, a TiVo and an internet radio in the lounge, a mobile phone, an iPod and a digital radio in the car, and countless other electronic gadgets around the house. You might even have an internet fridge.

If you're in your office, you'll see even more evidence of the fifth wave. At the very least, you'll have a fax machine or two, but if your office is a bit more up to date, they've likely been replaced by networked photocopiers that can scan, print and send. Computers take pride of place on every desk, and videoconferencing facilities are probably available in your boardroom.

Through your computer, phone and television, a whole range of new services are available. You can order a pizza online, buy a second-hand car, send flowers to friends on the other side of the world, and send mail or talk with colleagues for free over the internet.

Now picture your home or office without any of these items, and you'll get a sense of how much life has changed since the Intel 4004 hit the scene. But the fifth wave not only transformed our way of life, it also repainted the corporate landscape. In the 1970s, many of the companies that are household names today, such as Microsoft, Amazon, Google and eBay, didn't exist – these giants all emerged as a result of this technological revolution.

In the late 1990s, excitement around all of these technologies reached its peak. Investment in information and communication technologies, fuelled by business opportunities (and the fear of a massive 'Year 2000' glitch), reached a frenzy. Economists predicted that the traditional drivers of economic growth had changed, leading many to proclaim the dawn of the so-called 'new economy'. While these commentators might have got a little carried away, they were right about one thing: the fifth wave of innovation was in full swing.

But what made this wave so all-transforming? Like all waves, the information and communication revolution led to changes across the innovation spectrum – in technologies, markets and institutions. The changes in each area were impressive enough by themselves, but together, and reinforcing one another, these changes enabled spectacular innovation to occur.

Moore power

It is easy to identify the technologies that drove the growth of capacity in the information and communication industries during this fifth wave, but what is truly astonishing is the speed at which it grew.

In 1965, Gordon Moore, who a couple of years later would co-found the Intel Corporation, made what must have seemed a fanciful and bold prediction: that computing power would double every eighteen months, while costs would remain constant. He came up with 'Moore's law' after noticing that engineers had an amazing capability to double the processing power of the chips they were designing at a regular rate. His prediction has held true for the last forty years. Fanciful as Moore's law must have seemed at the time, computer processing power, as well as network bandwidth and data storage capacity, have all been following this extraordinary trajectory, and they look set to maintain it for at least the next fifteen.[1]

Moore's prediction was so bold because no other technology in history had ever grown in such a way before. Imagine an electric motor doubling its utility every eighteen months; if you started off with one horsepower – enough to run a small food processor – after thirty years you would have a motor with grunt in excess of one million horsepower – that's roughly equivalent to the output of a typical single-unit coal-fired power plant. And it wouldn't cost a single cent more than the original one-horsepower weakling.

So, from the original Intel 4004 chip, which had 2300 transistors, we now have its modern-day equivalent, Intel's Core 2 Duo processors containing over 291 million transistors.[2] Computer chips' adherence to Moore's law has given us the awesomely complex machines we have today, capable of tasks that could not even have been imagined when the 4004 was launched – predicting the weather, molecular modelling and even beating the human brain at chess.

This evolution has happened at a dizzying pace, and few companies have been able to keep up. One of those few is the International Business Machines Corporation, also known as IBM. While it is now one of the largest computer companies in the world, IBM is also one of the few such companies to have successfully 'surfed' multiple waves of innovation.

IBM began life in 1911 as the Computing Tabulating Recording Company, specialising in, amongst other things, punched-card equipment – the early ancestor of the modern-day computer. The company's first foray into the world of computers occurred in the 1940s, with the development of the Automatic Sequence Controlled Calculator, also called the Harvard Mark I. It was the first machine that could do complex calculations automatically, although it was painfully slow by today's standards. It could do addition in less than a second, but multiplication took six seconds and division problems took twelve seconds. The Mark I was also a tad unwieldy – at fifty feet long, eight feet high and weighing nearly five tons, it wasn't going to fit into anyone's pocket very easily.[3]

But it was at the vanguard of the technologies of the fifth wave. In 1952, the vacuum tube–based IBM 701 arrived, capable of executing 17,000 instructions per second, followed in 1964 by the first large family of computers, the System/360. Then in the 1980s the IBM Personal Computer was born, bringing computing into people's homes and offices for the first time. Its impact on society was so great that *Time* magazine elected the PC its 'Man of the Year' for 1982 – the first non-human to join the list.

IBM's spectacular trailblazing on the hardware side of things was mirrored by the Microsoft Corporation's success in software development. In fact, Microsoft probably owed its initial success to IBM. In 1981, IBM contracted Microsoft to develop the operating system for its personal computer – the result was MS-DOS. Not only was it used on IBM PCs, but as IBM PC clones started to emerge, Microsoft made its operating system available to them as well, and the company's fortunes soared.

Both IBM and Microsoft were perfectly positioned to take full advantage of Moore's law. As the processing capacity of its microprocessor doubled every couple of years, IBM continually brought out new versions of its chip, each more powerful than the last. This in turn enabled the creation of more powerful software, capable of ever more complex tasks. But as programmers pushed the envelope with regularly updated software, they began to demand more of the microprocessor – software and hardware became locked in an innovation arms race.

Jack in

The time soon came when personal computers were found in most offices and homes. But by themselves these mechanical brains have limited, albeit impressive, utility. One computer on a desk, isolated from the rest of the world, is hardly more than an amped-up abacus and typewriter. The real adventure began when computers and other electronic devices began to connect to each other.

Interestingly, Moore's law turned out to apply not only to the number of microprocessors that could be fitted onto a chip, but also to the speed of communications. Do you remember the first time you connected your computer to the internet? You were probably using what was then a 1200 Bd modem ('Bd' represents a *baud,* an individual signalling event in older modems). This meant that you could pump an astonishing 1200 bits down the phone line every second. But considering that it takes eight bits of information to form each letter in the Western alphabet, this was painfully slow.

In hindsight, it was amazing that we could do anything with this data stream at all. Twelve hundred bits per second is the equivalent of having a shower under a slowly dripping tap. However, we did manage to find useful things to send down this line, and a whole world of 'bulletin boards' sprang up, on which users could swap ideas and software.

Over the years, 1200 Bd turned into 2400 Bd, then 14,400 Bd (or 14.4 kilobits per second) then 28 kilobits per second then 196 kilobits per second.

These days, with Asymmetric Digital Subscriber Line (more commonly known as ADSL) and broadband to the home running at up to 100 megabits (millions of bits) per second, we are communicating at speeds inconceivable only twenty years ago. Rather than sending 150 characters per second over a phone line, we can now download a whole movie in minutes.

The combined effect of Moore's law on both computing power and communications ability created the technological leap for the last wave of innovation. But as we saw in the last chapter, technology alone was not enough to make the world embrace these new machines at such an astonishing rate; changes in institutions and markets were required as well. So what was happening to our institutions to complement these advances in technology?

The networked world

Imagine you and your best friend are the only two people on the planet with a mobile phone. This great new invention lets you contact each other anytime, anywhere. It's a neat setup for the two of you, as long as your friendship remains healthy and you continue to have plenty to talk about. But then two more of your friends acquire mobile phones – the usefulness of the device has just increased significantly, because you now have three times as many people to chat to while you're all out and about.

Then the technology really takes off, and soon everyone in your town, your city and your country has one. Everyone is connected, no matter where they are. The utility of the mobile phone has just taken a quantum leap forward. This is an example of a networked technology.

Robert Metcalfe, electrical engineer and co-inventor of Ethernet – a way of linking computers together into a local network – appreciated the importance of connectedness. Metcalfe's law states that the value and utility of a network are proportional to the square of the number of people using that network. Put simply, the more people connected to a network, the more valuable and useful it is. One person with a mobile phone can't call anyone; two people with mobiles can call each other; five people can make a total of ten phone connections; while ten people, each with a mobile phone, can create forty-five different connections.

Contrast this with a non-networked technology. Most things we buy are non-networked technologies. When you buy an apple, it doesn't really matter how many other people have also bought apples – you will probably get exactly the same enjoyment and nourishment out of your apple whether you eat it alone or in the company of apple-eaters around the planet.

Take another example. Antibiotics are one of the greatest discoveries of modern times and have saved many lives. If you have a serious infection, the value that a small pill can bring to you is enormous. But unlike a phone or email address, antibiotics have no network effect – the usefulness of this technology to you does not increase with the number of other people using it. With networked technologies, however, the number of other users makes all the difference.

Metcalfe's law also predicts a tipping point, at which time a network becomes so valuable that it is, effectively, irresistible. Once the network reaches a certain size, being left out carries a penalty – it's no longer a case of whether you can afford to be connected to the network, but whether you can afford *not* to be connected. When email first appeared, there were only a few users and the rest of the planet couldn't have cared less. But today those few who defiantly refuse to get email access are so greatly in the minority that they are undoubtedly inconvenienced as a result.

Be in it to win it

If you happen to collect broken laser pointers, you'd certainly have your work, or hobby, cut out for you. It's not the sort of collectable that would easily be found in antique stores, trading magazines or cereal packets. But software developer Pierre Omidyar managed to find perhaps the only collector of broken laser pointers on the planet when he first set up his online auction house in September 1995. Legend has it that Omidyar posted the item to test the site for the first time. Much to his surprise, the broken pointer eventually sold for US$14.83. The sale was so unexpected that Omidyar even contacted the buyer to make sure they knew it was broken.

It's not the strangest thing ever sold on eBay (the house where Bob Dylan grew up was successfully auctioned for around US$133,000), but it clearly illustrates that one person's trash is indeed another's treasure, and eBay is there to enable the transfer.

eBay is a great illustration of the black hole–like pull of a truly networked model. As the number of buyers hunting for treasure on eBay grows, so does the likelihood of a seller making a successful and profitable sale. This draws in more sellers, which in turn increases the likelihood that, if you happen to be a collector of weird or rare items, you will find someone selling them on eBay. So the more people using eBay, the more attractive it becomes. The rating system for buyers and sellers also creates trust and introduces a level of transparency that cannot be found in the systems of sale that eBay competes with, such as garage sales, trading magazines and auctions.

The tipping point for a network of any kind comes when people start to ask themselves: 'If I am not connected, will I miss out? If I don't have a mobile phone, will I miss out on calls? If I don't have a website, will my business suffer? If I am not on Facebook or MySpace, will people stop inviting me to parties?' The lure of these networks becomes greater and greater.

So it wasn't just about the stand-alone technology in the fifth wave, it was also the connections – the institutions that grew up in association with the technologies – that created massive shifts towards what has become known as the networked economy. With the internet at the core, businesses started to find new ways to connect, either with their customers (known as 'business to consumer', or B2C) or with other businesses (the even more catchy B2B). Networking became an institutional revolution.

Now every person and every business is connected by a sparkling array of new networked technologies, and those technologies are growing in capacity at a jaw-dropping rate. The question that still needs to be answered is: why? If the fire at the base of the fifth wave of innovation was an internet-powered Moore's law, and if it spread through the institutional use of networked technologies, what was its fuel? What was the market for all these new applications made possible by the internet? What made people want to fork over their hard-earned pennies in the first place?

The cost of doing business

To demonstrate the answer to this question, let's take a practical example. Put down this book and find a paperclip right now. That's right, do it. (We promise there is a point to this.)

How long did that take? Perhaps it was only a few seconds because you had one nearby. Perhaps it was a minute because you had to go downstairs and rifle through some papers. Or perhaps your inventive offspring have purloined every paperclip in the house to make a sculpture. To get one, you therefore had to walk or drive to the shops, perhaps stopping at an ATM on the way to withdraw some cash to pay for the paperclip, find a parking spot outside the shop, interact with the shop assistant to locate and buy your paperclip and then get back home again.

Each of those steps, and many other unseen ones, incurs a cost of some sort. There's the obvious cost of time – yours, the shop assistant's, the distributors who delivered the paperclip to the store, and so on. There's also the cost of petrol and parking if you drove, maybe a fee for withdrawing your money from the ATM, and the shop will also have costs of its own, such as its rent and utilities. Each of these costs – called transaction costs – will ultimately contribute to the overall cost of your paperclip search.

So what was the *true* cost of the paperclip? Most paperclips are so inexpensive that they only retail in boxes of 100, worth a fraction of a cent each. In fact, if we look at the time taken to fetch the paperclip, even if it only took one minute, at the minimum wage for many developed countries the cost of your time is still 100 times the cost of the paperclip, without the transaction costs taken into account.

Transaction costs are a fundamental force in business – exploration of their significance earned economist Ronald Coase the 1991 Nobel Prize in Economics. In his 1937 article 'The Nature of the Firm', Coase suggested that transaction costs could dictate the size and core business of a firm. For example, our one paperclip may only cost a fraction of a cent, but if you were to buy one paperclip at a time, the cost of your time would far outweigh the cost of the materials involved in creating the paperclip. By buying paperclips in bulk, you reduce the transaction cost of obtaining it to an acceptable level.

Coase took this idea one step further. He argued that some businesses actually came into existence in order to reduce transaction costs. For example, if a particular company had to buy boxes of paperclips and reams of paper individually, the transaction costs would be very high. As the company expands, these transaction costs continue to grow, to the point where the company is not simply doing its core business, but has become a massive paper and stationery logistics company. Finally, the point would be reached where someone high up in the company would decide it's not worth organising, stocking, maintaining and staffing these stationery activities, so they outsource those tasks to a business that exists specifically to manage stationery supplies for other companies. By outsourcing its stationery cabinet, the company has shifted away the transaction costs associated with this aspect of their business, and reduced its own size in the process.

The key decision in this outsourcing process is the cost of the transaction involved. Coase believed that firms would grow and shrink based on the marginal cost of an additional transaction. In other words, if you could find a firm out there that could provide the same service for a lower transaction cost, it would make financial sense to outsource the job to it.

In the late 1970s, the world was becoming more connected through the telephone, aeroplane and motorcar. Companies were growing bigger and supply chains were becoming more complex. With these increases in complexity and size, transaction costs within and between firms were starting to grow. The time was ripe for a technology to appear which would reduce these transaction costs, and capture the value of doing this.

Enter information and communication technologies. It may not have been very clear at the time, but with hindsight we can see that many of the technologies that have come to dominate in the information and communication wave have done so because they substantially reduce transaction costs. Sending an email is a lot cheaper and easier than writing a letter, getting to the post office and buying a stamp. Researching something online is many orders of magnitude faster than looking it up at the library, and online financial transactions – such as banking, buying and booking – incur a fraction of the costs associated with their earlier equivalents.

As the world was becoming more and more complex in the 1970s, transactions were becoming more expensive. For many items, the cost of transactions outweighed the cost of the object, particularly where maintenance was concerned – why go to the trouble of fixing something if you can just buy a new one for hardly more cost and with far less fuss? It was little wonder that when these transaction costs met the microprocessor, so much value was unlocked in the global economy.

In their book *Unleashing the Killer App,* Larry Downes and Chunka Mui suggest that the reason so many people chose to 'buy in' to new networked technologies was these transaction costs. Indeed, perhaps the last wave of innovation should not be known as the information and communication technology wave but rather as the 'transaction cost' wave, as this is where all the value came from. The companies that have done the best in this wave of innovation have done so precisely because they have identified a major transaction cost and effectively cut out the suppliers in the middle.

Computer scientist Jeff Bezos saw the incredible opportunities presented by the rapidly growing internet market. He identified an area that had not yet been successfully plundered in the hunt for dotcom dollars – online shopping. In July 1994 he founded Amazon.com , and one year later sold his first book online – the rather dull-sounding *Fluid Concepts and Creative Analogies: Computer Models of the Fundamental Mechanisms of Thought.*

Amazon now employs more than 20,000 people around the world and sells everything from books to shoes to power tools. Why is it so successful? Think about the effort, or the transaction costs, involved in the sale of a book. You have to find a store that sells the book you want and physically visit it to make your purchase. The store must employ a sales force, cover overheads such as rent, maintain a large enough selection of stock to attract most buyers but not so large that they are spending too much of their capital to hold books that might never sell ... the list goes on.

Amazon sells products with a minimum of transaction costs – you can shop from the comfort of your own home, the parcel arrives at your letterbox or your local post office, and you don't have the fuss of having to get cash out. On the business side, Amazon has drastically reduced the transaction costs normally associated with selling products, and because of its size it has enormous purchasing power. It has also built up a community of sorts, with buyers reviewing and rating books, which then encourages further participation and buying.

Bringing it together

Over time, simultaneous and harmonious changes in technology, institutions and markets underpinned the information and communication wave of innovation. Moore's law, by which the technologies of the fifth wave grew rapidly, supported Metcalfe's law, which drove the expansion of networks and linkages, which in turn enabled reductions in transaction costs, which then allowed for more investment in the technological drivers for Moore's law ... and so the cycle continued.

When you look closely at the successful companies of the fifth wave, you see that they all took advantage of each of these areas. Amazon's success depended on building a community around its book purchases, where people could provide ratings, reviews and recommendations. eBay reduced the costs of people transacting with one another, to the extent that it even created an opportunity for a company called PayPal to make these transactions cheaper; PayPal has since grown to be as large an enterprise as eBay itself. Even Microsoft took advantage of the network effects inherent in the sharing of documents with its Word software – once enough people were using Word to create documents, it became the de facto standard and anyone not using it was left out of the loop.

Many of the most successful businesses in the last wave fit this model. If this book was written thirty years ago we might have offered the following advice: 'Find the most expensive transaction cost that you can, and apply computing technology to it. Make sure that there is a networking component so that the more people who use it, the more useful it becomes.' If only we had a time machine...

The Wedding List Company is another example of using technology to reduce transaction costs in a networked environment. Wedding gift registries were commonly held by large department stores or speciality shops, which had its downsides. It meant the couple had to take time to visit and browse through the store, as did prospective present-buyers – a significant transaction cost. The store also had the usual overheads associated with a shopfront, and because of space and storage restrictions there was a limit to the range of products stores could offer.

Moving a wedding gift registry online drastically reduces many of these costs, while at the same time massively increasing the range of gifts to choose from. No more gift registry featuring the same tired old kitchenware and towels – now couples can request everything from a diamante-studded hammer to hot-air balloon flights.

Among each group of wedding guests directed to this site to find that special something for the soon-to-be-married, there is almost certainly a couple or two who will get a rush of blood to the head and decide to get married themselves. Having used the Wedding List Company to buy presents, they are likely to consider it for their own wedding, or if they haven't used it, they'll get referrals from friends. Thus the network expands organically.

A good online network can translate to a thriving online community, which creates a greater market for the original product or service. PayPal software engineers Jeremy Stoppelman and Russel Simmons set up Yelp.com in 2004 as an online directory of local businesses – if you were in San Francisco's Haight-Ashbury neighbourhood and were looking for a pizza restaurant, car mechanic or tattoo parlour, Yelp was intended to help you find it.

But the idea grew into something far more useful and interesting with the addition of users' reviews. Say you found a pizza restaurant in Haight-Ashbury and dined on the best pizza you had ever eaten. Most people would tell a few friends about their amazing pizza experience, but Yelp allows you to tell the whole world – or at least the part of it also using Yelp – by adding an online review to that restaurant's Yelp listing.

Yelp now features more than seven million local reviews written solely by its users, and that figure is growing steadily. To add to its appeal and reliability, Yelp has a similar system to eBay's whereby users can rate a review and reviewer, allowing regular, reliable and respected reviewers to build up a reputation and gain a following. And so, as the community around Yelp grows, more people use it and more businesses see the importance of paying to get a higher-profile listing on it – the network effect snowballs.

But then there are the technologies that don't fit this model. Remember the paperless office? As the personal computer came to dominate in homes and offices, many people began to envisage a world where every transaction would be conducted electronically and we would have no more need for sheets of paper, which ran the risk of being misplaced or destroyed. The paperless office was hailed as the 'office of the future' but strangely it has failed to eventuate. If anything, the advent of personal computers, printers and photocopiers has seen us use more paper than ever before. We make multiple copies of a document in the blink of an eye, we email a fairly unimportant memo to half the building and see just about every recipient print it out, and we print a page from a website to enable us to read it more easily.

When you examine the idea of the paperless office more closely, its failure thus far becomes less of a mystery. Having a paperless office wouldn't reduce the office's transaction costs very much. You might save on paper, and on the time spent walking to and from the printer, but these are relatively insignificant transaction costs in the grand scheme of things. Despite regular appeals for it to be adopted, the paperless office remains something of a mythical beast. But it may be that the paperless office was focussing on the wrong sort of efficiency for the fifth wave. It didn't reduce the cost of a transaction, only the resources involved. But as we shall see later, its day may be at hand.

The next wave

If the last wave of innovation was about reducing the transaction costs of purchasing an object, could the next wave be about reducing the costs of the resources used to make that object? It may be that we have finally reached the point where the transaction costs involved in ordering paperclips have been reduced so much that we are once again looking at the cost of the paperclip itself – specifically, at the cost of its raw materials.

Just as there was enormous scope to reduce transaction costs in the 1970s, today there is similarly massive potential to reduce waste. Now that the transaction costs of products have been reduced to a suitable level, the focus is shifting to how we might now create that product in a vastly more efficient and less wasteful manner. As the heady days of the boom in information and communication technologies draw to a close, a new wave of technologies is emerging that are challenging the domination of the Intel 4004 and its successors. The shifting of our focus from transaction costs to resource efficiency will have a profound effect on the world as we know it.

Chapter 3

Resource efficiency: the next great market

The death of a marine creature the size of a grain of sand many millions of years ago may not seem like a noteworthy event, but if you drive a car it's incredibly important. That creature's tiny corpse joined those of countless billions of its kind at the bottom of the ocean, and over many millennia, subjected to intense heat and pressure, it was transformed into the fuel for our modern world: crude oil.

Given the extraordinary conditions and length of time required to create crude oil, you would think we would treat this substance with respect and mete it out as carefully as another scarce and valuable resource – gold.

Think again. Despite the fact that this resource is finite and effectively irreplaceable, we consume crude oil with extraordinary inefficiency. Just fifteen per cent or so of the energy in the petrol you pump into your car actually goes into moving the car forward or powering accessories such as air conditioning.[1] A staggering eighty-five per cent of that valuable energy is simply lost as waste heat, pressure and noise.

The innovation trajectory of the petrol-powered motor vehicle has certainly been impressive; cars have come a long way since the first Model-T Fords rolled off the production line. But if, after nearly a century of innovation, the combustion engine can still only achieve an energy efficiency of fifteen per cent, perhaps we need a radical rethink.

Shai Agassi certainly thinks so. In fact, he's thinking about doing away with oil altogether.

Electric dreams

It's fitting that Agassi's story begins in Silicon Valley, where so many of the spectacular fortunes of the fifth wave were made. His story follows a familiar path. In his early twenties, this Israeli-born computer scientist founded several software companies, some in partnership with his father, that achieved considerable success in the Israeli marketplace, before they were acquired by software giant SAP for an impressive amount of money.

Agassi himself became integral to SAP, eventually rising to become president of the company's Products and Technology Group and sit on its executive board. In 2007, at the tender age of 38, Agassi was next in line for the position of CEO, but fate intervened and the much anticipated handover of the throne was postponed.

Two years earlier, Agassi had attended a World Economic Forum Young Global Leaders event where a challenge was issued to delegates to 'make the world a better place by 2020'. It prompted him to ponder how the world might run without oil. The question sat in the back of his mind as he continued his work at SAP, but finally the day came when that question had to be answered.

So, rather than stay on course to become the next CEO of SAP, Agassi resigned and founded Better Place, a company dedicated to a world of sustainable transportation. Why did he pass up the incredible opportunity to run one of the world's biggest software companies? Because he saw the development of sustainable transportation as an even bigger opportunity. The world was running out of secure oil resources, and was finding it ever harder to deal with the unwanted by-products – such as carbon dioxide – of burning the oil resources it did have. At the same time, the petrol-fuelled motor vehicle was very energy-inefficient; not only was much of that precious oil being wasted, but the ratio of 'unwanteds' produced to energy gained was far from ideal.

Agassi realised something that many other entrepreneurs are now starting to grasp. He saw that these key factors all pointed to an inescapable conclusion: that resource efficiency was the next great market opportunity.

Limits to growth

Since the Industrial Revolution – the first of Kondratiev's waves – economic growth has been intrinsically tied to resource consumption. Unfortunately, we live in a world of limited resources. Almost all the resources we consume in our daily lives, whether animal, mineral or vegetable, are finite, and for some of the resources the point at which we can no longer afford to harvest them is visible on the horizon.

This is hardly breaking news. Back in April 1968 the issue of runaway resource consumption already had Italian industrialist Aurelio Peccei and Scottish scientist Alexander King worried. They invited a select group of international professionals from diverse backgrounds to a meeting in Rome to discuss their concerns about the short-termism prevalent in international affairs.

Four years after this meeting, the group, calling itself the Club of Rome, released a report entitled *Limits to Growth*, in which it explored, to devastating effect, future scenarios of a world that pursued 'unlimited and unrestrained growth in material consumption in a world of clearly finite resources'.[2] These scenarios took into account variables such as population growth, industrial output, food per capita, non-renewable resource reserves and pollution output. One scenario in particular forecast 'overshoot and collapse' sometime in the twenty-first century.

The impact of this report was profound. While its authors went to some lengths to stress that they were not in the business of making predictions about the future, the conclusions of *Limits to Growth* were hardly optimistic:

> *If the present growth trends in world population, industrialization, pollution, food production, and resource depletion continue unchanged, the limits to growth on this planet will be reached sometime within the next one hundred years. The most probable result will be a rather sudden and uncontrollable decline in both*

population and industrial capacity.
[3]

Unfortunately, these predictions were pretty close to the mark, according to Dr Graham Turner, a senior research scientist with Australia's Commonwealth Scientific and Industrial Research Organisation (CSIRO). Turner has compared the scenarios the Club of Rome sketched out back in 1972 with thirty years of data collected since the report's publication, finding that the original predictions were dangerously accurate. 'From my view of this modelling, and factoring in some vague analysis of the likelihood of making really significant change ... it would be only logical and rational to think that we're going to hit the "overshoot and collapse" scenario,' says Turner.

The authors of *Limits to Growth* had expressed the hope that while all indicators at the time pointed to a dire outcome, there was still time and room to turn the *Titanic* around:

It is possible to alter these growth trends and to establish a condition of ecological and economic stability that is sustainable far into the future. The state of global equilibrium could be designed so that the basic material needs of each person on earth are satisfied and each person has an equal opportunity to realize his individual human potential.

Turner's research suggests we have failed to shift far enough from the original course, and are most closely following what *Limits to Growth* called the 'standard run' scenario – essentially, it has been 'business as usual' for the past thirty years. Just as the Club of Rome feared, we are still consuming our limited resources as though they were limitless.

The rate of our present growth might not seem that much. But even at only two to three per cent per annum, which is about the rate at which we are increasing our resource consumption, we double our consumption every thirty years or so. But so far so good, right? Well, yes – as long as you are not part of the last generation that is able to sustain this steady increase.

The British environmental biologist Sir John Lawton, chairman of the UK Royal Commission on Environmental Pollution, cites the well-known 'frog and the lily pad' problem to illustrate what is so alarming about our consumption rate. 'The frog has to have a lily pad and it has to have some open water – it needs both those resources,' he says. 'The lily pad population doubles every week, and so after a certain number of weeks the frog is perfectly happy because half the pond is water. The next week, it's all gone.'

This is the sting in the tail of exponential growth, says Lawton. 'Exponential growth has a nasty habit of dumping stuff very, very rapidly on you, and then the question is, is there time to respond?' Which leads us to the dreadful question: when will it be too late for us?

A world of limited resources

Sir John Lawton says it's difficult to talk definitively about the limits of natural resources because technology can almost always find a solution, albeit at a price. However, there are some very rough estimates on how much of our 'pond' is left. The conservation biologist Stuart Pimm, in his book *The World According to Pimm,* suggests that humans consume around forty per cent of global plant production, using it for food, animal fodder and timber, around a third of total production from the world's oceans, and sixty per cent of accessible freshwater run-off.[4]

These figures are all rather alarming, especially when our exponential growth in consumption is factored in. Lawton believes the situation with water is the direst, particularly for landlocked nations that don't have access to seawater for desalination.

This is prompting some countries to go to great lengths to manage their scarce resources. Despite being an island, Singapore has a water problem – specifically fresh water. It's not that it doesn't have enough of it during the wet season, but it is short of places to store it. However, instead of simply building a land-based reservoir, Singapore took the unusual approach of turning an entire bay into a reservoir by means of a dam across the Marina Channel. The reservoir created by the Marina Barrage is expected to supply around ten per cent of the island nation's water.

You would be hard-pressed to find anyone on the planet ignorant of the fact that our most precious resources, particularly oil and water, are becoming scarcer. This dawning awareness is gradually prompting a measure of thriftiness in some institutions and sectors of the population – for example, the implementation of water restrictions in a large number of nations around the world, and moves in a variety of countries to penalise owners of larger, less fuel-efficient cars with higher taxes or fees. An entire town in Australia even recently voted to ban plastic water bottles because they represent such a waste of oil.

Yet these efforts are being undermined by the rampant inefficiency in how we use resources, not just in car engines, but across many industries. Most of us are blissfully unaware of the true resource cost involved in the manufacture of the many products we use and consume in our daily lives.

Hidden consumption

It is not just the consumption we can see which is growing, but the consumption we can't see as well. For example, if you live in the developed world, there's a good chance that you're currently wearing a pair of denim jeans, or at least that you have a pair in your wardrobe. Denim is made from cotton, which is an extremely water-intensive crop, accounting for around 3.5 per cent of global water use for crop production. Huge amounts of water are used to irrigate cotton crops, and to bleach, dye and wash the fabric to give us the denim we know and love. A single pair of adult's jeans requires around 11,000 litres of water to produce – but we don't see a drop of it.[5]

It's a similar story with many familiar foodstuffs that grace our plates. According to the Water Footprint Network, production of a single slice of wheat bread uses a whopping forty litres of water, which works out at around 1300 litres per kilogram of bread. This inefficiency is widespread. The water efficiency of agricultural products (measured in dollars per megalitre) can range anywhere in the scale from US$500 for wheat to US$10,000 for out-of-season cherries.

Tap water can almost be thought of as liquid electricity – a huge amount of energy goes into processing and transporting it – yet we use it to wash down our driveways and flush our toilets. And it's not just water we need to worry about. Resource inefficiency is endemic to modern life, and the true resource costs of most products are hidden. The computer this book is being written on consumes a certain amount of energy, and as we feel particularly strongly about environmental issues, we have purchased this energy from a 'green' energy supplier. But what about the energy and resources consumed in the manufacture of this computer, its shipping, storage and sale, and finally its delivery to the end-user?

Think of the computer chip, upon which the fortunes of the fifth wave were built. It's a very small object, yet the amount of resources that go into its production is far out of proportion to the chip's actual size. By employing a process called life-cycle analysis (or life-cycle assessment), we can get a sense of exactly what resources are consumed to produce a single computer chip.[6]

The first ingredient to consider in a computer chip is crystalline silicon. As the second-most abundant element in the Earth's crust, there's no shortage of silicon – but it still needs to be mined, probably through an open-pit mine. Silicon mining causes slightly less environmental damage than the mining of many other minerals, but it's still a huge operation involving heavy machinery, consumption of fuel, output of carbon dioxide (CO_2) and other pollutants, and transportation of the resulting ore. The silicon is then purified via a chemical process that requires it to be heated to between 950° and 1150°C, which burns up lots of fuel.

The purified silicon is then sent to a factory to be cut into wafers, but only forty-three per cent of the silicon actually ends up in the wafer – the rest is waste. Some of this waste is used for alternative products, such as solar panels, and some might be recycled, but a proportion of it is simply discarded.

Chemicals are then used to etch circuits into the surface of the silicon wafer. This part of production is quite water-intensive, as purified water is used to wash the chemicals off the wafer. Because the waste water contains chemicals, it can't be flushed back into the waterways but has to be carefully treated. Production of a single 32-megabyte computer chip, weighing around two grams, requires more than thirty litres of water. Throughout this process, the computer chip must be kept in an absolutely spotless environment, as a single impurity can ruin the chip. Keeping dirt out of the manufacturing process requires a lot of filtration and a lot of energy – approximately 1.5 kilograms of fossil fuels per chip.

Then there's packaging, transportation and distribution, throughout which other resources are consumed; a proportion of those are wasted. All in all, for each gram of silicon computer chip, 630 grams of fossil fuels are used.

It's a bit like an iceberg – just a small fraction of the resources consumed are actually visible to the end-user, or even quantifiable. But below the surface lurks a huge, unseen and dangerous volume that, as maritime history has already demonstrated so clearly, is what gets us in the end.

Supply and demand

The English city of Newcastle upon Tyne has contributed much to the world: Newcastle Brown Ale, the famous 'Geordie' accent, a fine football team, and coal – lots of coal. Its coal reserves were once so significant that the phrase 'carrying coals to Newcastle' is a well-worn idiom that describes a pointless or foolhardy endeavour – an illustration of the fundamental principles of supply and demand. There was no demand in Newcastle for an external source of coal, and no profit to be made in attempting to import coals to this area. The only historical record of anyone defying this was an American entrepreneur who, in the eighteenth century, managed to make an unlikely profit from delivering coal to Newcastle during a miners' strike.

This was the only recorded case – until now. As the once-plentiful coal reserves around Newcastle dwindle, the miners must delve deeper and deeper to extract them. As the mines become deeper, the cost of mining goes up, and for the first time in modern history it has become more economical for Newcastle to import coal than to dig it out. The port of Tyne now receives regular shipments of coal from Russia and South America, while other ports around the United Kingdom also receive coal from North America and even from as far afield as Australia.

As dwindling resources get consumed at an accelerating rate, the notion of supply and demand becomes important. Completely aside from the long-term concern about exactly what we are going to put into our fuel tanks or make our computers from when we've run out of oil, there is a simple economic reality: as supplies becomes scarcer but demand continues to grow, prices go up. While the lucky few who happen to be holding those resource cards are likely to do very well indeed, the vast majority of the population will face a scenario in which the resources we take for granted today are prohibitively expensive tomorrow.

Wasteful practices or unsellable production?

For an environmentalist, the wasting of eighty-five per cent of the petroleum energy pumped into a fuel tank is a travesty because oil is such a precious and limited resource. But the individual who just paid to pump the petrol into his or her fuel tank has quite literally seen eighty-five per cent of their petrol money vanish in a puff of exhaust fumes. Likewise the supermarket that has to throw out huge amounts of unsold fresh produce at the end of each day, and the telecommunications company that has to invest in expensive heat dissipation technologies for its computer server rooms, and the farmer who must spend a fortune on irrigation water, some of which will be lost to evaporation – all are spending money on something that does not generate any return.

What is especially shocking is exactly how much of the raw material used for a product ends up as waste. In their book *Natural Capitalism,* Paul Hawken, Amory Lovins and Hunter Lovins suggest that the amount of material extracted, used and ultimately disposed of is approximately twenty times the body weight of each consumer in the American economy annually. On a global scale, this equates to a resource flow of around a half-trillion tons per year. And yet within a year, less than one per cent of that resource flow remains as a durable product; ninety-nine per cent of the original raw materials are waste.[7]

This is another way of looking at waste. Waste is unsellable production – something you have produced but cannot sell. Waste is the dead zone of profit. It can be seen as the ultimate transaction cost. Just as the game plan of the fifth wave was to reduce the transaction costs of selling to zero, the game plan of the sixth wave will be either to reduce this unsellable production to zero or, conversely, to find a new way to make money from it.

This is a huge opportunity. Just as Jeff Bezos built his online empire of A mazon.com by substantially reducing the transaction costs associated with buying books and other products, many budding entrepreneurs of the sixth wave are launching their fortunes on the back of what is essentially rubbish.

Waste not, want not

So what do you get when you join the market driver of increasingly costly inputs with huge inefficiencies? A very big incentive to create resource efficiency.

Resource efficiency is at the heart of the sixth wave of innovation, as we shift away from harvesting resources that are plentiful and cheap to managing resources that are scarce and valuable. Resource efficiency can come about through a number of pathways: from companies saving money by reducing packaging, to individuals who cut their electricity bills by using better insulation. If you can think of a way to do the same thing with fewer resources, you are starting to get at the market driver for the next wave.

The pressure has been mounting for some time to improve the efficiency with which we consume resources, and this is particularly evident in the car industry. If you bought a car ten years ago, you probably didn't think too much about how many litres of fuel it would guzzle per hundred kilometres of driving. Car advertisements focused on a car's appearance, performance, power, luxury, interior design ... just about everything except how fuel-efficient it was.

Now every other car advertisement talks about fuel efficiency. This is partly because many governments around the world now require all new motor vehicles to be sold with a fuel consumption label. But with petrol prices rising, consumer preference is also shifting away from large, powerful, fuel-hungry cars and towards smaller, lighter, more efficient – and, therefore, ultimately cheaper – models. It's a shift that some companies in the automobile industry are struggling to catch up to, as evidenced by the recent turmoil in the US auto industry, where giants such as General Motors and Chrysler filed for bankruptcy.

It's not only cars that are being made to account for their resource consumption. Most whitegoods sold today are labelled with stickers showing their energy-efficiency ratings. As households grapple with their environmental conscience and their utility bills, a process of natural selection is driving these goods towards greater resource efficiency. Could we see a new Moore's law emerge that relates to resource efficiency? Just as the processing power of a computer chip increases exponentially, perhaps we'll soon see a similar effect in the resource efficiency of refrigerators and washing machines.

Money for nothing

The expression 'where there's muck, there's brass' is particularly well suited to the sixth wave. While originally intended to mean that there is good money to be made from dirty jobs, the emerging industry of landfill mining is taking the phrase much more literally.

Professor William Hogland likes to joke that he doesn't plan to leave anything for his children to inherit as he wants to live life to the full; besides, if they can work out the technology, there will be more than enough riches left behind in landfill. As Professor of Environmental Engineering and Recovery at Sweden's Linnaeus University, he should know. Hogland has been interested in landfills since the mid-1990s, when countries such as the United States were beginning to appreciate the value buried in their piles of waste.

Initially, much of the interest in landfill mining was focused on extracting material that could be used as extra fuel for incinerators. Hogland says a landfill can contain as much as seven or eight megajoules of energy per ton of waste – even more if there is a higher plastic content. 'In Asia, you have landfills that have a plastic content of twenty per cent or more, and then you have a good fuel, or even if it's good quality you can recover the plastic'.

And this is the second benefit of landfill mining – extracting material that can be reused or recycled. 'You can get good economy by excavating for recovery of materials,' Hogland says. 'For example, in a car-wrecking landfill, you have car demolition parts, you have electrical waste, you have rubber from the cars and so on. Maybe it can be an economic benefit to excavate such a landfill.' Other materials such as aluminium and copper can also be extracted, and some companies are looking into the possibility of excavating such landfills to get at these resources. However, it's still early days for landfill mining, and many are waiting for the prices of the resources to increase to a point where it becomes economically viable to extract the materials within.

Landfill is also a potent source of methane gas, which is produced when organic material breaks down in anaerobic conditions. Methane is a particularly nasty greenhouse gas – it is twenty times more effective at trapping heat than carbon dioxide – and landfills are one of the biggest manmade sources of methane. Many people are now realising the potential energy to be gained by tapping this gas and using it to generate electricity. US company Waste Management Inc.'s Palmetto Landfill in South Carolina supplies sixty per cent of the electricity needs of the nearby BMW manufacturing facility by using the methane gas to power the plant's turbines. This has meant a saving of around 92,000 tons of carbon dioxide each year.

Another surprising resource being extracted from landfill is dirt. 'After fifteen to twenty years, in most landfills, forty-five to sixty per cent is soil,' Hogland says. This soil can be used either to cover over the landfill after the waste has been extracted, or it can be used simply as soil. 'I have shown in some of my studies that the quality is quite okay for most of the parameters – it's almost compost standard.'

Sometimes brass can be found without having to dig through the muck at all. Platinum is considered more valuable than silver or gold because of its rarity – estimates suggest that all the platinum ever mined would barely fill a room measuring 7.5 metres on each side.[8] The amount of new platinum reaching the market each year is less than five per cent of total gold production – King Louis XV of France once declared it the only metal fit for a king for this very reason. In 2006, just 239 tonnes of platinum were traded globally, but it is an indispensable resource in the modern world – just about every motor vehicle on the road today carries platinum inside its catalytic converter.

Were it not for the catalytic converter, the air in our cities would be considerably more polluted than it is. A catalytic converter is a device that reduces the amount of noxious vehicle emissions by converting the more harmful substances in car exhaust, such as nitrogen oxides, carbon monoxide and unburned hydrocarbons, into less harmful substances, such as nitrogen, oxygen and carbon dioxide. The main component of a catalytic converter is a honeycomb structure whose inside surface is coated with a metal catalyst that helps the conversion of the bad stuff to the less bad stuff. The most common metal catalyst used is platinum.

The platinum is only present in trace amounts, however even trace amounts add up very quickly when there are so many cars in use: more than half of the platinum mined around the world ends up in catalytic converters. Unfortunately, like so many other precious metals, platinum is becoming increasingly rare, and this has seen its price skyrocket in recent years.

Dr Hazel Pritchard is particularly interested in platinum. As an exploration geologist at the University of Cardiff, she has devoted considerable time and energy to hunting for raw metal deposits in natural geological environments. But when searching for platinum, she kept coming up empty-handed. That was until someone told her platinum was being thrown out of cars, but not in the sense of chunks of it being tossed out of the window – it was coming out of the exhaust.

So off she went with a brush and pan and literally swept the roads and roundabouts of Cardiff, collecting road dust. The results astounded her. 'I was amazed to discover how much there was,' she says. 'Platinum is very rare and very precious – you can mine it at five parts per million – and when I swept the roundabouts in Cardiff, with a little bit of mineral processing I was getting two parts per million.' That's a dramatic increase on the natural background levels of less than one to three parts per *billion*. And it wasn't just on the roads. 'I've done work in the UK, and [there] it should be almost non-detectable if it's natural, but because we're introducing it into the environment in such an amazing way it's almost in everything, even estuary mud on mudflats.'

There's even a relatively easy way to mine this unexpected bonanza – street-sweepers, which Dr Pritchard says suck up thousands of tonnes of road dust each year. But there is a catch: the platinum in road dust has a different mineralogical form to that found in platinum ore, and the technology doesn't yet exist to extract pure platinum from this form. But with the price of platinum rising steadily, and the importance of catalytic converters in keeping the air in our cities relatively clean, it's only a matter of time before someone works out an economical way to mine this manmade source.

An even smarter way to harvest waste would be to catch it before it reaches the garbage bin or landfill. Consumers throw away a phenomenal amount of perfectly good food before it even makes it onto the plate. Food-Wise, an Australian organisation, estimates that the equivalent of one in five bags of groceries is abandoned, which amounts to more than $5 billion of food waste annually.[9] Another study looked at the amount of food produced in the United States and compared it against the amount of food actually consumed – the difference was 150 trillion kilocalories a year in waste, or about forty per cent of the US food supply (up from twenty-eight percent in 1974).[10] More than half of this is fresh food: the soggy lettuce at the bottom of the crisper, the uneaten slices of ham way past their prime or the unused jar of pasta sauce, for example. Most of this organic matter ends up in landfill, where it generates enormous amounts of methane.

While the bulk of food waste comes from households, a considerable portion also comes from retailers such as supermarkets. In the United Kingdom, retailers are blamed for generating 1.6 million tonnes of food waste each year – around twelve per cent of all the food wasted in the UK.[11]

This costs both retailers and the environment, so in 2000 a UK consultancy, the Organic Resource Agency (ORA), with funding from a variety of sources including major supermarket chains, undertook a project investigating whether organic, biodegradable waste from supermarkets could be managed in a cost-effective, sustainable and environmentally sound way through composting. The result was

GROWS – Green Recycling of Organic Waste from Supermarkets.

GROWS was about 'closing the loop', according to waste-management expert Dr Jon Pickering, senior consultant at ORA and project manager for GROWS. 'It's not rocket science,' Pickering says. 'What made sense was the fact that a farm could take waste material from supermarkets and convert that into a resource that could then be used to improve or maintain soil fertility, which could then be used to improve the crops or livestock, which could then be sold back to the supermarket.' It's simple, yes, but Pickering says there were plenty of unanswered questions. Would supermarkets be willing to invest in separating out and depacking compostable waste? How could they collect the waste from supermarkets? What would be the best composting method? Would there be any problems with weeds? Was it possible to have biodegradable packaging suitable for a range of composting methods, including home composting? And, perhaps most importantly, would the compost actually make a difference?

Over a sixteen-month period the project collected fruit, vegetable and cut-flower waste from twenty-seven UK supermarkets, composted it using three different methods, and then distributed the compost to a large organic farm. The result? 'The only issue is economics; there are no technical issues,' says Pickering. 'If somebody can make a profit out of doing it, they will overcome the technical issues.'

Pickering and colleagues found it was eminently possible for supermarkets to segregate organic waste and even de-pack the waste in the store, although there was a certain degree of reluctance and the project required a strong champion at management level to ensure it was done and done properly. With some effort, the team also managed to get contamination of the waste stream down to a fraction of a percent, albeit via a labour-intensive manual process.

The three composting techniques – open outdoor composting, covered outdoor composting and a compost bin-like system – were all successful at converting the waste, with the main difference being the amount of emissions produced, the cost and the ease of management. In field trials, all three types of compost generated increased soil fertility and grain yields, with no weed or disease problems.

The biggest challenge was collecting the waste. 'The collection service was the key barrier; it proved to be the deal-breaker in the end, because the difficulty was we were trying to collect small individual volumes of waste from a number of different stores,' says Pickering. 'Most waste-collection services operate best with high density. Here, we had a collection area across most of the south of England.' In an ideal world, each town would have its own waste-collection service, and the waste would be composted and used locally.

Even this was not an insurmountable problem, yet the final conclusions of the GROWS project suggested that it was an idea whose time had not yet come. 'At the time we did the work, landfill costs were still considerably cheaper than the cost of collecting and disposing of waste through composting,' Pickering recalls. 'Effectively, the scheme couldn't be rolled out because there wasn't sufficient financial incentive at that time.'

However, since the completion of the GROWS project, landfill tax has increased in certain parts of the UK; suddenly, the ideas behind GROWS are starting to look a lot more attractive. Rather than simply discard leftover food products as waste, an incentive has been created to encourage the efficient use of this resource.

Tapping into people power

Resource efficiency also extends beyond oil, water or food to more intangible things, such as skilled labour. We waste hundreds of thousands of hours of skilled labour every day.

For example, if you make a comment on someone's blog or fill in a web form, there's a good chance you'll be asked to type a series of distorted numbers or letters shown in an image. The aim of this apparently pointless exercise is to determine if you are a human, or an automatic program such as a spambot – the idea being that an automated program would be unable to translate the distorted figures as accurately as a human brain. These CAPTCHAs (Completely Automated Public Turing test to tell Computers and Humans Apart) are widespread security measures on the World Wide Web.[12] Their use is so widespread that, every day, more than 100 million of these CAPTCHAs are typed in by people around the world. It takes only a second or two, but that soon adds up to a lot of person-hours and a lot of text.

A group at Carnegie Mellon University discovered that they could put all of this character recognition to another use. Instead of having people transcribe random letters and numbers, each CAPTCHA sequence was a word or two scanned from an actual publication – most often an ancient text or rare book. By getting people to transcribe this fragment of text, they are using the CAPTCHA to digitise ancient texts and rare books. Their system, called reCAPTCHA, can achieve ninety-nine per cent accuracy and is now deployed in more than 40,000 web sites. It has already transcribed over 440 million words.[13]

Without even knowing it, people are doing two things at once – verifying their humanity and transcribing an ancient manuscript. The Carnegie Mellon group has managed to tap in to a valuable human-resource stream without the resource itself even being aware it is being exploited. But there are other ways of making efficient use of available labour resources, such as by thinking beyond geographic boundaries. One unusual way is to hide the resource behind a computer interface.

Two centuries before the IBM supercomputer Deep Blue was going head-to-head with chess champion Garry Kasparov, another mechanical mind was gaining fame by taking on the notables of its time – including Benjamin Franklin and Napoleon Bonaparte – at chess. And it was winning. The Mechanical Turk was a chess-playing machine built in the eighteenth century and exhibited as a wondrous automaton.

But the Mechanical Turk was a fake, operated by a human chessmaster hidden inside the cabinet upon which the automaton sat. And as much as eighteenth-century audiences might have wanted to believe that an automaton was capable of performing a task as complex as playing chess, the technology to enable this was still a long way off.

Such is also the case with the Amazon Mechanical Turk. If you weren't aware of the inner workings of this website, you could be forgiven for thinking that a mechanical brain was doing the thinking and delivering the results. But, as with the original Turk, the explanation is people power.

The Amazon Mechanical Turk is a crowd-sourcing website. Crowd-sourcing is like outsourcing, but instead of contracting the work out to a particular person or company, you throw the opportunity out to a crowd of willing and able people, and then cherrypick the best result. Say you're a health-related charity, for example, and you need to update your list of medical organisations that operate within your state. You could get an in-house staffer to do the task, but even if they were earning a basic wage, this simple but time-consuming task would still cost your organisation a reasonable sum of money.

So you consult the Amazon Mechanical Turk – a web service provided by Amazon.com. You post your task on the website and offer to pay a certain amount of money – say, five dollars – for someone to complete the task. Anyone anywhere in the world can choose to undertake your task, and you only pay up when you're satisfied with the job they've done. Five dollars might seem a token amount of money to pay, but many of the tasks posted require no specific skills other than a reasonable command of English, and there is no shortage of people around the world with this ability who would be willing to earn five dollars for an hour or two's work.

Innovation Exchange and InnoCentive operate on similar crowd-sourcing principles, but the stakes and rewards are much greater. At Innovation Exchange, companies and organisations offer tens and even hundreds of thousands of dollars to anyone who can help find a solution to a tricky problem, such as helping to develop a new alcoholic beverage brand, identifying ways a hospital system can cut costs without negatively affecting services, or brainstorming ways to increase donations to combat child poverty. At InnoCentive, the tasks require an even greater level of skill in the sciences, mathematics and business, with challenges, such as finding a biomarker to measure disease progression in Lou Gehrig's disease, offering a reward of US$1 million to the successful applicant.

Crowd-sourcing is a global response to a global talent war. With a dwindling supply of skilled workers, companies have to think outside the four walls of their buildings. They are becoming entities bound only by the limits of information and communication technology's reach, as the drive towards resource efficiency generates some very clever opportunities to unlock human capital.

Sifting the wheat from the chaff

We might be running out of oil, fresh water, food and skilled workers, but one thing we are definitely not short of is information. In fact, we're drowning in it. If you type 'Sixth Wave' into a search engine, it returns more than 900,000 hits in an astonishing 0.16 seconds. Among the results are websites for music groups, economic theories (not the same as the premise for this book, however), battles and politics – all clearly valid search-engine results for the search term. The previous information and communication technology wave has given us the incredible boon of the internet, which has placed an unprecedented universe of information at our fingertips, but unfortunately much of it is irrelevant, unreliable or completely unusable.

Information is one of the few resources that is actually increasing in quantity, but unfortunately not in quality. The challenge of this resource is to sift the wheat from the chaff – to filter the few bits of useful, relevant data from the maelstrom of rubbish. Innovations that improve the efficiency of this resource are becoming highly sought after, and one company that has stepped up to the plate in spectacular fashion is Google.

While Google is a creation of the fifth wave, where it emerged as one of the giants of the 'new economy', it looks set to make an equally impressive performance in the sixth wave. The Google search engine's ability to return the most relevant results for a search query has seen it rise above its competitors to dominate the search engine market. This is impressive enough, but Google's position as the most widely used search engine has opened up another unexpected opportunity to further exploit this information resource.

Visit www.google.com/trends on any given day and you get a sneak peek into the global zeitgeist for that day. Google Trends examines trends in search terms – for example, on the day this chapter was being written, the untimely shooting death of an American football quarterback dominated the search terms, occupying three of the top ten entries. Many people were also looking for information about Independence Day fireworks in the United States. These are pretty innocuous search terms, and it may be hard to see what use can be made of this facility other than as a minor amusement.

But say you and your family have been feeling a little under the weather, suffering coughs, aches and fevers. You suspect it might be influenza, but rather than go to the fuss of visiting your local doctor for something that might turn out to be nothing more than the common cold, you go online and search for 'influenza symptoms'. If there is an outbreak occurring in your part of the world, there's a good chance quite a few other families will be doing the same thing.

By analysing trends in search terms related to influenza, Google was able to map the ebb and flow of influenza searches. It found a close relationship between the number of people searching for influenza information and the number of people actually suffering from the infection. This is a very useful discovery for health authorities, as a spike in influenza-related searches could serve as an early warning sign of influenza outbreaks, allowing health authorities to implement preventive measures and education programs.

It's not just influenza that health authorities want to get the jump on. Recent experience with SARS and the H5N1 and H1N1 influenza viruses have highlighted the importance of early-warning systems for diseases that have the potential to spread rapidly. But it's the viruses we haven't yet encountered – those that haven't yet made the jump from animal to human, and from human to human – that we most need to be alert for. Imagine how differently the emergence of HIV might have played out if Google Trends had been available.

Wolfram|Alpha takes this information efficiency one step further. Rather than being a search engine, Wolfram|Alpha is an 'answer engine'. The idea is that the website collects and curates hard facts about anything and everything, then structures that data in such a way that it can 'compute whatever can be computed about anything'. So if you want to get a factual answer to a question – for example, you might want to know the height of the Empire State Building in New York – you ask the answer engine. Instead of referring you to a webpage that contains the answer, it will actually provide you with the answer directly. Its founder intends it to become a single source of knowledge that can deliver factual answers to factual questions.

Thinking outside the oil barrel

With resource inefficiency so widespread throughout modern life, a wealth of opportunity exists for anyone who can find ways either to minimise the inefficiency, to maximise the efficiency, or to come up with a completely new solution.

This is just what Shai Agassi, the founder of Better Place, has done. With the car industry staring down the barrel of rising oil prices and a dwindling supply, Agassi recognised that the inputs fuelling the global car industry were only going to decrease over time, and that what we needed was an alternative model.

That alternative is the electric vehicle, but beyond simply changing the machinery, Agassi's vision transforms the way we view the car. When you make the decision to buy a petrol-fuelled vehicle, you might think that you are simply buying a sleek piece of machinery. But effectively, you're not only paying the cost of the technology, you're also signing an implicit contract with the petrol industry for the time you own and drive that car. Given the car's inefficiency, the highly undesirable by-products that affect not only the environment but also your own health, and the fact that we are steadily paying more and more for the petrol, it's hardly a fair contract.

Agassi's plan is to change the nature of that contract to one that more closely resembles those that accompany our mobile phones. When you get a new mobile phone, you generally don't actually pay for the phone itself – not up-front, at least. You sign a contract with a mobile phone provider, whereby you get a phone for a fraction of the full cost of the technology, but in return for that discount you subscribe to use that mobile phone company's services for a period of time.

So how does Agassi's idea work? First, there are the cars. Electric vehicles are nothing new – in fact, they predate petrol- and diesel-fuelled vehicles, and for a while they actually outsold them. But the cheapness of the combustion engine, and perhaps also pressure from the growing petroleum industry, forced the electric engine to take a back seat – until now. Electric cars are experiencing a renaissance, and not a moment too soon. Being battery operated, they substantially reduce, or totally eliminate, almost every pollution type associated with the combustion engine (depending, of course, on the source of the electricity).

Firstly, and most obviously, these cars do not produce any exhaust. Secondly, the engine operates almost silently, drastically reducing noise pollution. Thirdly, the use of regenerative braking to recharge the battery means no brake pads, which in turn means no brake pad dust – another pollutant associated with petrol-powered cars. Fourthly, electric vehicles also have far fewer moving parts than conventional cars, so less maintenance is required and there are fewer parts to be replaced, therefore fewer resources are consumed and less waste is produced. Electric cars are also far more efficient than petrol cars – the cost per kilometre of running an electric vehicle can be up to seventy per cent less than that of a petrol-powered car.

The cost of running an electric vehicle is largely the cost of the electricity to power it. Agassi's vision would see charge points located anywhere that you might park your car – your garage, in office car parks, public car parks, roadsides, and anywhere else that a car might legally be left. You simply pull up, plug in and your battery will recharge as you go about your business. And it almost goes without saying that the electricity will come from renewable sources such as solar and wind.

A typical battery would last around 160 kilometres. If you're going on a longer trip, you'll need a battery-swapping station. Just as petrol stations currently dot the landscape, so will the battery stations. When your battery is getting close to exhaustion, you pull in to one of these stations and an automated system will remove your old battery and replace it with a fully charged one – without you even having to get out of the car. You drive off with your fresh battery, and the depleted one is recharged at the station for use in another car. And to make sure that you have enough juice for your trip, or at least to get to the next battery-swapping station, your car's on-board computer will work like an intelligent satellite-navigation system, accessing real-time traffic information to work out the best route to your travel destination, and it will guide you to charging spots or battery-swapping stations when your power is low.

It might sound like a fanciful scenario, but Agassi's vision is well on the way to becoming reality. His idea caught the attention of Israeli president

Shimon Peres, who has backed Agassi in implementing the world's first nationwide charging network for electric vehicles. The first 'plugged-in' parking lots were unveiled in December 2008, and more are to come. Better Place is also taking the first steps to develop similar networks in Denmark, Australia, Canada, Hawaii and California.

The stone age didn't end because we ran out of stones

Shai Agassi has a saying: 'The stone age didn't end because we ran out of stones.' In the same way, the oil age will not end because we run out of oil. But it is becoming more and more apparent that the way we currently use our resources is inefficient and wasteful, and as they become scarcer, there is an increasingly compelling reason to do more with less.

Agassi's story is not unique. Many other entrepreneurs are seeing the market in resource efficiency, the market driver for the sixth wave of innovation. However, there is more to resource efficiency than simply improving how a resource such as petrol is used. Resource efficiency is only part of the story of the sixth wave; there are institutional factors at play as well. Another massive opportunity exists, and it's one that Agassi and others like him are well-placed to take advantage of – the increasing trend towards putting a price tag on waste. As we'll see in the next chapter, the world is far less tolerant of the old way of managing waste, unwanted by-products and pollutants. The old habit of simply 'throwing it over the back fence', in a manner of speaking, is no longer acceptable, and waste has become a costly affair. And as we have seen in previous waves, where there is cost there is opportunity.

Chapter 4

Capturing true costs: evolving institutions

If the polar bear is the poster animal for global warming, the orangutan is the face of deforestation. And what a great face it is – expressive dark eyes, orange fur, fabulously long limbs and jowls to put any overweight patriarch to shame. It's hard to put a value on such a face, but it is very easy to put a value on the trees that this unique creature resides in.

In 2005, an estimated US$64 billion worth of timber was harvested from the world's forests, largely for industrial use.[1] One-third of the world's forests are used primarily for the production of forest products, which amounts to a staggering three billion cubic metres of wood harvested in 2005 alone. That doesn't even include illegal logging – an estimated seventy-three per cent to eighty-eight per cent of all logging activity in Indonesia is illegal.

Rainforests also face a huge threat from palm oil, as global demand for this food product and biofuel soars.

In the face of such a monster profit incentive, the orangutan stands little chance, and the forests of Sumatra and Borneo that it calls home are being decimated at a depressing rate. The United Nations Environment Programme estimates that by 2022, at the current rate of deforestation, ninety-eight per cent of the natural rainforests of Sumatra and Borneo will have been destroyed, felled mostly for their timber.[2]

While most of the industrialised world has been tearing its hair out about the plight of the orangutan, one man saw an opportunity. Australian internet entrepreneur Dorjee Sun saw that there was a new crop on the block, and one lucrative enough to give logging and palm oil a serious run for their money: carbon.

Sun's vision is of a future where it is more profitable to preserve rainforests than it is to chop them down. Instead of profits coming from the sale of rainforest timber or palm oil, they will come from the sale of those forests' ecosystem services – in particular, their ability to absorb carbon dioxide. His company, Carbon Conservation, specialises in brokering the sale of carbon credits from *avoided* deforestation – the preservation of forest that would otherwise have been logged – to firms looking for ways to offset their carbon emissions. But these transactions are about far more than the buying and selling of carbon credits.

What Sun has succeeded in doing is exploiting the value of something that was previously valueless, at least to the people chopping down the trees. Thanks to carbon trading, it is now more profitable to see the forest for the trees, so to speak.

As we saw earlier, waves of innovation are powered not only by significant market opportunities, but also by massive institutional changes. Putting a price on carbon is just one example of the sweeping institutional changes that are accompanying the sixth wave. These institutional shifts are revolutionising the way we see our natural environment, and they're putting an economic value on things that have never been valued before.

But to understand why this change is happening now, we need to look at the drivers of the change. The first and foremost reason is that our natural resources are facing an unprecedented threat: us.

The population bomb

In the early 1970s, three leading American scientists were locked in a passionate debate about what had made the greatest contribution to humanity's impact on the environment. The environmentalist and biologist Barry Commoner argued that advances in technology since World War II had been the most significant. However, the entomologist Paul Ehrlich, author of the famous 1968 book *The Population Bomb,* and the physicist John Holdren, now director of the White House Office of Science and Technology Policy, maintained that humans' environmental impact was an equal function of population, technology and consumption.

The debate led to the creation of what is now known as the IPAT formula, which states that humans' impact on the environment (I) is the product of population (P), affluence (A) and the impact of technology (T): or I=PAT. An increase in just one of these parameters, therefore, increases our environmental impact.

The world's population is now approaching seven billion people, and we are expected to reach nine billion by 2040. While a few nations, such as Japan, Russia and Italy, are actually experiencing negative population growth, most of the rest of the world's populations are growing, albeit at a reduced rate in recent years. Growth rates are particularly high in the Middle East, South and South-East Asia, Latin America and Sub-Saharan Africa. Every new person is a new mouth to feed, a new body to clothe and keep warm, a new consumer of resources and a new producer of waste.

According to Ehrlich and Holdren,[3] population growth has a disproportionately negative effect on the environment; a one per cent increase in population doesn't just mean a one per cent increase in environmental impact. Firstly, they argued that pollution has a synergistic effect, so that different pollutants can interact and enhance their individual effects on humans and the environment – the whole effect is greater than the sum of the individual effects. Secondly, they suggested that a threshold for pollution exists; below it, the environment can cope with the pollution, but above it, the system breaks down. For example, 500 people living around a lake might empty their raw sewage into the lake, and the lake's natural processes might be able to break down that sewage without harm. But if the population around that lake increases to 700 people and the lake is unable to cope, the environment begins to suffer. Thirdly, the law of diminishing returns means that as populations increase, we are demanding more and more production from our agricultural land; eventually, we resort to fertilisers and chemicals in an attempt to wring more from less. Whichever way you look at it, population growth is bad news for the environment.

However, the 'A' of the equation – affluence – is also on the increase. To be affluent is to have an abundant supply of wealth, whether in monetary form or in commodities. In economic terms, it's an indication of the level of consumption per person, and is often measured as gross domestic product (GDP) per capita. So while China may be experiencing a population growth rate of around 0.6 per cent per year, its GDP per person is increasing at more than ten per cent per year. Likewise, India's population is growing at 1.5 per cent annually, but its GDP per capita is increasing by 7.7 per cent, and with increased wealth comes increased consumption.[4]

This means that people in China and India are now more than ever able to 'keep up with the Joneses', and they're wasting no time in doing so. Accoutrements such as luxury cars, flat-screen televisions and mobile phones – commonplace accessories in a Western lifestyle – were once well out of reach for the masses. But as these products become cheaper, and as the average individual wealth of people in these nations increases, they are becoming more attainable. An increasingly aspirational and rapidly enlarging group of consumers is buying up big-time, and the result is greater consumption of resources and more rapid creation of waste.

Finally, we come to the 'T' variable of the IPAT equation, which is a measure of the role technology plays in meeting the needs of these increasingly affluent and enlarging populations. This particular variable is less set in stone than the other two, as it is a measure of the *intensity* of resource consumption, meaning that better, more efficient technologies can lead to fewer resources being consumed. With a low T-value, a more efficient society is able to get more from less, consuming fewer resources to meet the same or increasing needs and generating less waste from the same amount of production. Unfortunately, as we have seen, in our modern society technology has not yet been able to keep pace with increases in population and affluence.

According to the IPAT formula, our planet is therefore rocketing towards a destructive future. And we are feeling the first rumblings of its approach.

The price of success

More than two thousand years ago, the town of Linfen, Shanxi Province, held the distinguished position of the capital of all China. These days, according to the Blacksmith Institute, an organisation dedicated to tackling pollution, Linfen holds the more dubious distinction of being one of the top ten most polluted places in the world.[5] The air in Linfen is thick with fly-ash, carbon monoxide, sulphur dioxide and nitrogen oxides spewed out by the hundreds of nearby coalmines, steel factories and refineries. Drinking water is so polluted and scarce that it is rationed even in the provincial capital. More than half the well-water in the province has been found to be unsafe for drinking due to its high levels of arsenic, and the local children have high rates of lead poisoning.

Linfen is one of the Blacksmith Institute's 'Dirty Thirty' – a list of the most polluted places on earth. But this list is just the tip of the iceberg. The World Health Organization estimates that air pollution is responsible for more than two million premature deaths each year, the majority of which occur in developing nations.[6] In stark contrast to the dwindling supplies of resources such as oil, water and platinum, one thing the modern world has in abundance is pollution. Carbon monoxide and dioxide, arsenic, cadmium, chromium, lead, mercury, PCBs, volatile organic compounds, radioactivity, sulphur dioxide, particulates ... the list of culprits is depressingly long.

To address its high-polluting reputation the government of Linfen created the 'Blue Skies, Green Water' initiative. This program has involved the closure of hundreds of mines and factories that were creating the waste products – particularly hard-hit were small businesses who couldn't afford the investment required to become cleaner.

These closures came at a cost, reducing Linfen's GDP by almost US$300 million in 2007 according to state statistics.[7]

Waste does not just affect business – many of the costs of pollution are borne indirectly by society. For example, climate models are predicting changed weather patterns as a result of climate change, with more droughts in certain parts of the globe and more violent weather in other parts. To a consumer, this translates into more expensive goods (as water becomes scarcer), higher taxes (as the community pays for damage to the environment and society) and higher insurance premiums (as the risk of disaster increases).

The cost of waste

Wednesday 14 April 1999 started out just like any other autumn day in Sydney. The weather was calm, with the presence of a weak cold front along the coast and some showers in the nearby Blue Mountains suggesting a mild thunderstorm later in the day, but nothing to cause any concern. At around 4.25 pm, a thunderstorm did take shape south-south-west of Sydney; it quickly developed into a more substantial storm that dumped minor hail on the coast, earning it a reclassification as 'severe', before it moved out to sea. At this point, there was still nothing to alarm staff at the Australian Bureau of Meteorology enough to issue a severe weather warning.[8]

But at around 7.30 pm all hell broke loose. The storm turned inland again, and gained such force that it has since been labelled a 'supercell' – a rare but extremely severe type of storm characterised by a massive rotating updraft. It struck the coast just south of Sydney Airport, before moving north to unleash its full fury on the city's eastern suburbs and central business district. The area was battered by golf ball–sized hailstones, some as large as nine centimetres in diameter, and wind gusts of up to eighty-five kilometres per hour.[9] A second, smaller thunderstorm passed over the same area not long afterwards, adding insult to injury with further hail and massive downpours.

It proved the costliest natural disaster in Australian history. As well as claiming one life and causing injuries to around fifty people, the storm damaged 24,000 homes, 70,000 vehicles and twenty-three aircraft. The final insurance bill was $1.7 billion – the largest ever on the island continent but, including the uninsured damage, the final cost may have been as high as $2.3 billion.

Tony Coleman was the chief risk officer for Insurance Australia Group (IAG), which incurred claims totalling more than $300 million following the 1999 hailstorm. This single event represented one-quarter of IAG's total weathe-rrelated claims in New South Wales for fifteen years,[10] which understandably raised concerns about whether such a devastating storm could happen again. IAG began looking at computer modelling of the atmospheric factors that might lead to the development of hailstorms in the Sydney region, and came to some disturbing conclusions.

'That led straight into the issue of climate, and discovering that the incidence of hailstorms in Sydney was pretty closely connected with the temperature of the ocean off Sydney in the twenty-four to forty-eight hours before the hailstorm,' Coleman recalls. 'Of course, if you have the temperature of the ocean more often warm than not because of global warming, the probability of that happening increases.'

Understanding weather-related risk is core business for the insurance industry, as weather-related payouts constitute the largest proportion of total catastrophe losses by insurers. Unexpected and catastrophic events like the 1999 Sydney hailstorm are potentially disastrous for the insurance industry. They have, therefore, a very strong interest in knowing as much as possible about what lies ahead.

IAG's research suggested that major storms like Sydney's are simply a taste of what's to come, and they're not the only ones to have reached this sorry conclusion. In 2004, the Swiss reinsurance giant Swiss Re forecast that the economic costs of natural disasters were going to increase dramatically due to climate change, doubling to US$150 billion per year by 2014. This translates to US$30–$40 billion in insurance claims, and the money has to come from somewhere.[11]

'There's no doubt that insurance premiums are starting to rise because of this – insurers have no choice,' says Coleman, now a company director with an interest in risk-management and climate-change issues. 'You could describe them as the economic barometer of climate change.' Insurers are now looking at the world through climate-change-coloured glasses, and are examining anything that might be affected by climate change a lot more closely, particularly when it comes to what Coleman calls 'high-risk geography'.

'Climate change is changing flood risk, and generally increasing it,' he says. 'If you've got a piece of land that is more exposed to flood risk than not, what's beginning to happen is you don't get cover at all for flood risk, or if you do, you get charged separately and extra for it.'

It also means insurance premiums are starting to factor in the long-term sustainability of ventures in a climate-changed future. For example, in the wake of devastating bushfires in the Australian state of Victoria, the Royal Commission charged with investigating the causes of and responses to the fires is looking at tighter 'bushfire-proofing' regulations for any structures rebuilt in fire-prone areas. There's little doubt that insurers will be paying a lot more attention to these issues as well. Climate change might be perceived by some to be a problem to be faced in the future, but insurance companies are already feeling the pinch. They are bringing the issue of climate change into the here-and-now.

So what can be done? It has become increasingly clear that something has to change in order for us to reduce the negative impact we are having on the environment. In response to this challenge, we are seeing some significant institutional changes, which all have something in common: they are putting a value, or cost, on things that have never been valued before.

Early institutional changes: regulating waste

Since the dawn of humankind, we have had to deal with waste. Our earliest ancestors created middens, or dumps, for their refuse and other unmentionables. These were sited away from living spaces, signifying a very early recognition that some of the by-products of day-to-day life need to be disposed of safely, away from where people live and conduct their business. Unfortunately, that innate wisdom has gone astray somewhere along the line. We now have the bad habit of dumping our unwanted waste in our own backyards. Since the 1970s, institutional reforms have attempted to address this problem with a very simple and visible solution: fine the polluters. It's the first of a raft of institutional changes to benefit the environment, and one that states very clearly that waste is no longer someone else's problem.

Finding oil in all the right places has made the fortunes of countless individuals and companies. But leaving oil in the wrong places has cost some of those companies a significant part of those same fortunes. When the *Exxon Valdez* oil tanker struck Bligh Reef in the pristine waters of Alaska's Prince William Sound in 1989, it spilled around forty million litres of crude oil into the surrounding waters. The spill wasn't the largest in history, but was arguably one of the most environmentally devastating. The courts certainly thought so, awarding US$5 billion in punitive damages against Exxon (although on later appeal the amount was reduced to around US$500 million).

Colonial Pipeline also found itself on the wrong end of an oil slick, with the accidental release of around 5.5 million litres of oil and other petroleum products from a number of its pipelines in several American states. For those transgressions, the company was slapped with a US$34 million fine in 2003 by the US Environmental Protection Authority – at the time, the largest civil penalty paid by a company in EPA history.

But it's not just oil that has cost polluters dearly. In 2007, American

Electric Power paid a US$15 million penalty to the EPA for violations of the Clean Air Act at its sixteen coal-fired power plants around the country; it also faced a bill of more than US$4.6 billion for installation of pollution control equipment. And in 2005, DaimlerChrysler paid US$94 million for defective catalytic converters on nearly 1.5 million Jeep and Dodge cars.

Fines of these sizes, even for companies that make as much profit as these do, provide a pretty powerful incentive for them to change their practices and avoid repeating the environmental transgressions. This institution of regulating waste is one example of how the world's governments are responding to the impending environmental crisis. Waste is no longer something that can be accidentally, or surreptitiously, chucked over the back fence.

But pollution disasters such as the *Exxon Valdez* oil spill are very much the worst-case scenario and, thankfully, are relatively rare. The more frequent challenge is to discourage the slow, steady release of unwanted by-products into the environment. Here again, though, when money talks, industry listens.

Internalising externalities

A more sophisticated way to address the issue of waste is found at the very heart of the world's market economy. At the same time that regulations were being brought in to fine polluters, economists were making the point that waste is an example of what is known in their field as a 'market failure'.

Market failures occur when something stops market forces from doing what they do best – allocating resources efficiently. A market failure can occur when the market does not take into account all the costs or benefits associated with a product or service, particularly when those costs or benefits are a side-effect of a transaction. These side-effects are called *externalities.*

An externality is created when something you do or something you make indirectly affects someone else. It can have a positive or negative effect, but the key things are that the effect is inadvertent – it isn't part of the main 'transaction' – and that you don't incur any additional costs. Second-hand smoke is an example of an externality, because while smokers are happily transacting with their cigarettes, those around them (as well as the smokers) are suffering the negative consequences of that transaction through the release of carcinogens and other toxic by-products. Externalities can also be positive. The building of a new public transport link to a suburb will almost certainly raise the value of houses in that location, much to the delight of those living there.

Whether positive or negative, the very existence of an externality is considered an indication of market failure because costs or benefits have 'slipped through the cracks'. A smoker is not directly reimbursing bystanders for the cost of his or her second-hand smoke, while the builders of the public transport link are not benefitting from the increased profits of landowners in the area.

Market failures can cause significant problems for the community as a whole; if producers don't have to shoulder the costs of their waste, there is no incentive for them to control or limit it. And when you look at things from a market-failure perspective, it becomes clear that the community, and the planet, have been subsidising industry for many years by bearing the costs of environmental degradation.

But all that is changing. As awareness grows of the environmental impact of waste in its many forms, industry is at last being forced to 'internalise' these externalities. What was once free now comes with a price tag. But other than simply fining or taxing companies for releasing waste and other toxic emissions, is there a way to put a real price on all these externalities?

Cap and trade

Economics has a solution to this, and it is known as *cap and trade.* In order to put a price on waste, you first cap the total amount of waste that can be emitted, then you allow companies to trade their permission to emit waste with one another. Because the emission of waste will be more valuable to some businesses than others, traders will quickly settle on the monetary value of these 'emissions permits'.

Cap and trade systems have already proven effective at encouraging companies to keep their externalities under control. In 1999, the European Union issued a directive on landfill waste, which aimed to slash the amount of biodegradable municipal waste ending up in landfill by sixty-five per cent by the year 2010. The United Kingdom's response was to introduce the notion of landfill 'allowances', where a single allowance represented a fixed amount of biodegradable municipal waste that could be sent to landfill. They then put a cap on the number of allowances granted to waste authorities and set up a scheme to allow authorities to buy and sell allowances. Authorities could also bank allowances for future years if they managed to reduce their landfill below their annual allowance, or borrow up to five per cent against future years.

The Landfill Allowance Trading Scheme was launched in April 2005.

The first year saw more than twelve million tonnes of waste sent to landfill – a large amount. However, in the second year of the scheme, waste authorities around the country managed to reduce landfill to 20.5 per cent below their allowance, and in the 2007–08 year, there was a further 8.4 per cent reduction.

By allowing trading between the waste emitters, the market was able to put a dollar figure on a tonne of waste. Suddenly value could be created by reducing the amount of waste that was produced; there was a financial incentive to find better ways of reducing waste – through more efficient sorting and recycling, for example.

Effectively, what a cap does is introduce *scarcity* in the amount of waste that can be released. This creates a limited supply, so demand and supply meet at a certain price. Cap and trade can also work for less 'solid' forms of pollution; it has proved successful in reducing the amount of sulphur dioxide pollution in the United States.

Acid rain is an environmental ailment unique to an industrialised world. It first came to general attention in the mid-nineteenth century, when people began to notice that forests downwind of industrial complexes were dying. The phenomenon was further explored in the United Kingdom in 1852 by the Scottish chemist Robert Angus Smith, who found a link between the atmospheric pollution over Manchester and acid rain. The issue was largely ignored for another century, until the rising acidity in some lakes and streams brought it back into the scientific consciousness.

The particular chemical culprits responsible for acid rain are sulphur dioxide and nitrogen oxides – often described by those in the business as 'SOx and NOx' – both of which are released as by-products of coal combustion. These react with water, oxygen and other substances in the atmosphere to form acidic compounds that fall to earth either in a 'wet' form (as rain or snow) or a 'dry' form (as gases or particles).

In 1991, a program established by the US Congress to investigate the problem of acid rain filed its first report, and the news wasn't good. The report showed that five per cent of lakes in New England were acidic, which had devastating consequences for their aquatic life. Six per cent of those lakes were so acidic that many species of minnow could not survive in them, and two per cent were no longer able to support populations of brook trout.

To tackle the problem, the US government created the Acid Rain Program. The first phase targeted coal-fired power plants in twenty-one states, requiring them to cut their sulphur dioxide emissions to a combined total cap of 8.7 million tons by January 1995. Phase two, launched in 2000, expanded the program to cover more than 2000 power plants. The end goal was to wind back annual sulphur dioxide emissions to pre-1980 levels, representing a cut of around ten million tons annually.

So that's the cap, but what about the trade? Each power plant was issued with a certain number of emission allowances, with a single allowance representing one ton of sulphur dioxide. The number of allowances given to each plant was decided by their sulphur dioxide emission levels during the 1980s. At the end of each year, each plant tallies up its sulphur dioxide emissions and works out whether it has come in under its allowance or over it. If the plant has exceeded its allowance it must buy more allowances, but if it's managed to reduce its emissions enough, it can sell its excess allowances.

By allowing trade between emitters, the market was able to put a price on sulphur dioxide. Those who could reduce their emissions more cheaply were able to attract investment. In effect, those that could make the easiest reductions in emissions did so, selling their permits to those who could not. Some even banked on the sale of these permits to invest in further reductions.

By January 1995 – the end of the first phase – power stations had cut their emissions to 5.3 million tons of sulphur dioxide, which represented a drop of around fifty per cent compared to 1980 levels.[12] The environmental benefits are already being seen. Levels of wet sulphate, an important component of acid rain, have dropped by as much as fifty per cent in most of the north-east and midwest of the United States. And the entire program has ended up costing around one-quarter of the EPA's original estimates.

But by far the most ambitious cap and trade scheme is yet to be implemented, and it's the most important of the lot. It will set limits on the compound that poses perhaps the greatest environmental threat of all: carbon dioxide.

A low-carb(on) diet

As a former chief economist of the World Bank, Sir Nicholas Stern knows a thing or two about economics. In 2006 he wrote a report on the global economic impact of climate change, in which he argued that climate change was 'the greatest and widest-ranging market failure ever seen'.

According to Stern, climate change can be seen as the ultimate externality. He explained his position in a lecture to the UK's Royal Economic Society in 2007:

> ...if we emit greenhouse gases, that causes damages to other people, and at present we don't pay for the damages which we cause to others ... That is an externality, it is a market failure, and we called this the biggest market failure the world has ever seen because greenhouse gases are involved in everybody's activities and the impacts of climate change hit everybody, and they hit everybody, potentially, on a really major scale. [13]

Carbon dioxide is the ultimate environmental bête noire, and the primary agent of climate change. The 1997 Kyoto Protocol was the first significant step towards dealing with this monster of an externality. The international agreement was linked to the United Nations Framework Convention on Climate Change, which in 1997 set legally binding targets, or 'caps', on emissions, for the four worst greenhouse gases – carbon dioxide, methane, nitrous oxide and sulphur hexafluoride – in an attempt to avoid the worst climate-change scenario.

Under the Kyoto Protocol, signatories commit to reducing their emissions of these four greenhouse gases, as well of hydrofluorocarbons and perfluorocarbons, by a certain percentage of their 1990 emission levels. These reductions are to be achieved between 2008 and 2012. For example, the United Kingdom committed to an eight per cent decrease, while the Russian Federation agreed to stabilise its emissions to 1990 levels and Australia was allowed an eight per cent increase on its 1990 emission levels.[14] The 2009 Copenhagen Accord took a small step further with the United States, India, China and many other nations not already bound by Kyoto committing to limit temperature rise from climate change to two degrees Celsius above pre-industrial levels. The accord also took steps towards setting up an emissions verification system, and industrialised countries pledged, by

2020, to provide an annual US$100 billion a year in aid to poorer nations to assist in climate change mitigation and adaptation measures.

The Kyoto Protocol also incorporated an important element of flexibility through the creation of an emissions trading scheme. It's a classic cap and trade arrangement, in that it provides an economic incentive to nations to reduce their emissions. The cap means that there are a limited number of carbon emissions permits, which are then traded by nations that have a demand for carbon dioxide emissions. Demand and supply meet at a certain price; the cost of emitting a tonne of CO_2 or its equivalent in methane, nitrous oxide or sulphur hexafluoride.

This has led to the formation of a carbon market, through which carbon 'credits' (each one equivalent to a tonne of carbon dioxide) are bought and sold. Those who release more than their allowance of greenhouse gas must buy excess credits from those who have released less. And it's certainly a lucrative institution. In 2006, the global carbon market was worth an estimated

US$30 billion. One year later, it had more than doubled its value to US$64 billion and then nearly doubled again the following year to an astonishing US$126 billion.[15]

As well as allowing trading in these emissions, the Kyoto Protocol also incorporated something it called the 'Clean Development Mechanism', which recognised the inevitable fact that many of the most cost-effective emissions reductions were more likely to be found in the developing world than in developed nations. Put simply, it was likely to be cheaper for industrialised countries to invest in reducing emissions in the developing world than it would be for them to reduce their own emissions. So industrialised countries are permitted to invest in certain developing-world projects; in doing so, they earn emission reduction credits.

There are regulations about what projects may be invested in; one key criterion is that the emission reduction or removal project would not otherwise occur, were it not for the incentive created by the credits. Projects currently on the list include the building of wind farms in China, the upgrading of a composting facility in Delhi, and the establishment in Brazil of plants that convert landfill gas to energy.

These new systems pose an interesting legal challenge, as they represent legally binding international agreements that are implemented at national and local levels through domestic law and policies. It's a unique situation that has given rise to yet another new institution – the practice of climate change law.

Martijn Wilder, partner and global head of law firm Baker & McKenzie's Climate Change practice, was there from the very beginning. 'I had done lot of international law and a lot of environmental law, and I was quite involved with WWF and others, and as a result of that I had pretty much a commitment to the area,' he says. 'I really wanted to do international environmental issues, and this was the issue, or was going to be the issue, as far as I was concerned.'

While the cap and trade system provides a financial carrot to encourage industry to reduce its emissions, equally there needs to be a stick, and that's where the law comes in. 'It's critical for trying to resolve climate change,' says Wilder. 'If you don't have laws that reduce emissions, and do it properly, then the planet cannot reverse dangerous climate change.'

Climate-change law differs from environmental law in that while environmental law is involved in environmental planning, environmental impact assessments and pollution licences, climate-change law is essentially about regulating emissions through market mechanisms. 'If you want to restrict companies, they're not going to do it of their own free will, so you need a stick to force them to do it,' Wilder says. 'That stick is either putting a tax on carbon so the more you pollute the more it costs you, or it's making carbon a right with scarcity so people trade it and value it.'

And that's exactly why people like Dorjee Sun are able to do what they do. In March 2008, Sun and his company, Carbon Conservation, succeeded in convincing banking giant Merrill Lynch to invest US$9 million to protect 750,000 hectares of rainforest in the Indonesian province of Aceh. It was a hard-fought victory for Sun, who faced an uphill battle to convince the corporate investors of the value of protecting tracts of rainforest as carbon sinks, but he was aided by Aceh's governor, who declared a ban on logging in the area.

The plan is relatively simple. 'By avoiding deforestation you are avoiding emissions, by avoiding emissions you earn carbon credits, and by earning carbon credits you can monetise them,' Sun says. So Merrill Lynch, for its US$9 million investment in protecting the forests of Aceh, is actually buying around 3.3 million carbon credits,[16] which it can then sell for upwards of US$400 million. It's a win–win situation for the bank, but what about the rainforest?

Merrill Lynch's US$9 million will contribute to measures that not only reduce legal and illegal logging, but also promote reforestation and involve local communities in sustainable management of the forest. These include employing former Free Aceh rebels to police the forest for illegal logging activity, and helping villagers to farm crops such as rubber, cocoa and timber sustainably rather than with the slash-and-burn approach that has been so devastating to the area. The idea is that instead of the locals being paid to chop down the forests, they will be paid to nurture and protect them.

The project was the first REDD (Reducing Emissions from Deforestation in Developing countries) program to meet the Climate, Community and Biodiversity Alliance's rigorous standards for land-based carbon projects. But for Sun it's only the first step. He believes carbon is simply the first cab off the rank when it comes to pricing externalities, and that soon everything from biodiversity to clean air will be managed and nurtured in the same way.

And he's right, because in the sixth wave we will not just stop at pricing waste – we will begin to put a value on all environmental resources, from water to biodiversity, soil to wind.

Ecosystem services

The value of a litre of water is pretty clear – it relieves thirst, washes dirt away and nourishes plants, to name just a few of its many functions, and it is easy enough to put a price on some of these. But how do you measure the value of a tree? Its most obvious value is as a source of timber, and in that form it is easy to price. It might also provide food in the form of fruit, nuts and seeds; the value of those can be determined by a trip to your local supermarket. Many would argue that a tree also has aesthetic value; plant nurseries around the world would no doubt agree.

But a tree's most important job is actually its least valued, by humans anyway, which is strange because our very survival depends on it. Trees perform a complex function that can best be described as 'servicing the ecosystem'. Trees absorb carbon dioxide and emit oxygen, they stabilise soil cover and prevent erosion, they help prevent salinity by lowering the water table, they provide a home for other living things, they provide shade from the summer sun, shelter from the wind – the list goes on.

Ecosystem services, as they are known, are any goods or services provided by the ecosystem that are of value to humanity. These goods and services are incredibly diverse but can be classified in four categories: provisioning services, which supply consumables such as food, water or timber; regulating services, which influence systems such as climate, disease and water quality; cultural services, which benefit us in aesthetic, spiritual and recreational ways; and supporting services, which contribute to nutrient cycling, soil formation and fertility, through photosynthesis.[17] But how can you put a dollar figure on these services?

One place where the true value of a tree has come to be appreciated is the Catskill Mountains, in the north-east of the United States. The Catskills are a beautiful part of the country. In autumn the vibrant red, orange and yellow foliage is reflected in the still lakes, and the clear air carries the first hints of winter bite. Whether your idea of fun is fly-fishing, camping, hunting, hiking or just warming your toes in front of a roaring fire in a log cabin, the Catskills have something to please everyone.

Far more importantly, though, the Catskills are a life-support system for the nearly nine million people who live in and around New York City. Ninety per cent of the 1.3 billion litres of water consumed every day by the residents of this metropolis comes from six reservoirs, all located in the 4100 square kilometres of the Catskill Watershed.[18]

As well as supplying the bulk of New York City's water, the Catskill Watershed supplies water so pristine that since 1993, the city has been exempted by the US EPA from having to build water-filtration plants to treat its water. Filtration is costly, both in energy and resources, so this ecosystem service is not to be taken lightly.

In 1996, however, New York City was in danger of losing this special exemption because the EPA felt that not enough was being done to guarantee the safety and quality of the watershed. To protect the catchment, an agreement was negotiated between city, state and regional authorities, the EPA, and environmental and agricultural organisations. It included the acquisition of nearly 150,000 hectares of environmentally sensitive land; adoption of regulations governing systems such as storm-water control and residential septic treatment; careful management of agricultural activities within the watershed; and a stream management and restoration program. All these activities are estimated to have cost New York City around US$1.4 billion dollars.

This sounds like a lot of money but, given the services that the Catskill Watershed provides, it was only a drop in the ocean. The EPA has estimated the water-treatment systems that would otherwise have been needed to filter the Catskills water would have set the city back US$6–8 billion, not to mention the ongoing maintenance costs.[19] In view of that, investing US$1.4 billion in the natural capital of the Catskills to maintain the quality of the ecosystem services it provides seems like a sensible financial decision. What's more, the trees of the Catskills offer a range of other ecosystem services that haven't been priced, from the carbon dioxide they absorb to the tourist dollars they attract. But the valuation of its water-filtration services alone has finally meant that the ecosystem services provided by the Catskill Watershed are no longer taken for granted. And New York City is not the only place considering investment in watershed preservation as opposed to water-filtration plants – more than 140 other cities are also coming to appreciate the financial wisdom of securing ecosystem services.[20]

But it's not enough simply to put a dollar figure on an ecosystem service. As New York City found out, just as important is to define who owns it in the first place.

Property rights

'The inherent logic of the commons remorselessly generates tragedy.' So it was that American ecologist Garrett

Hardin rather morosely set the scene for his now famous 1968 dissertation in *Science* on the tragedy of the commons.[21] The 'commons' he refers to is a shared and free pasture used by a group of herdsmen to graze their cattle. That's fine and dandy when the number of herdsmen, and therefore the number of cattle, are kept within a certain limit by disease, wars, poaching and other mechanisms of attrition. A balance is maintained, such that the demand on the pasture as a resource is kept at a sustainable level. But the day comes when better health, less conflict, better security and general improvements in living standards mean there are more herdsmen, and a lot more cattle. At this point, the balance breaks down and the predicted tragedy rears its ugly head.

Because no one owns the commons, each herdsman will, naturally, try to graze as many cattle as possible on the pasture before the resources are depleted. And because no one owns the pasture, no one person is liable for the full penalty of the pasture's degradation. So it becomes a race to see who can be first to exploit what's left of the resource. As Hardin describes it:

> *Each man is locked into a system that compels him to increase his herd without limit – in a world that is limited. Ruin is the destination toward which all men rush, each pursuing his own best interest in a society that believes in the freedom of the commons. Freedom in a commons brings ruin to all.* [22]

Unfortunately, 'freedom in a commons' is exactly the state of affairs that exists with many of our natural resources, from water to biodiversity. These things are shared by all of us, whether we collect the water from a well or tap, whether we use that water to irrigate fields or wash our cars, whether we clear a native forest to farm cattle or sell the timber or whether we hunt animals for food or fun. But in many parts of the world, no one 'owns' the water, the forest or the animals – and so no one has ultimate responsibility for them. And because of that, we are beginning to see Hardin's predicted 'ruin to all' – our water is becoming steadily more scarce and polluted, our native forests are dwindling and our biodiversity is disappearing.

The 'tragedy of the commons' is a tragedy because there are no property rights – no one owns the pasture, so no one person cares for it or protects it. Property rights include the right to access a resource, the right to use and benefit from it, the rights to control the use of it and exclude others from using it, and the right to transfer on or sell ownership of that resource.

An illustration of the effects of a lack of property rights is a slum. In a slum, none of the occupants are able to own the land they live on, trapping them in a cycle of poverty, deprivation and degradation. There is no security of tenure, and because of this there is no incentive for people to invest in the future of the slum. In fact, the Peruvian economist Hernando de Soto Polar maintains that the main reason why the poor of the developing world remain poor is because they have no property rights. Without them, they have no collateral to secure loans that would allow them to buy better homes or build new businesses and thus escape a subsistence living.[23]

When property rights are put in place, there is an incentive to invest in protecting the resource that is owned. Those who receive value from resources and those who own resources can start to trade, and the value of those resources starts to become evident. In some cases, resources such as water can even be traded between different users.

We have been living in the equivalent of an environmental slum, where the environment and the ecosystem services it provides are neither valued nor protected. But all that is changing as we begin to take ownership and responsibility for our natural resources. In drought-stricken Australia, debate has been particularly fierce over the question of water property rights. But how do you decide property rights over something that falls freely from the sky or rises up from the ground, and that runs unencumbered over the earth?

Professor Mike Young, director of the Environment Institute at the University of Adelaide, says it's more useful to think in terms of entitlements, rather than property rights, when it comes to water. 'Well-designed entitlement systems are built around notions of sharing what's available, and running an independent and separate process to determine how much is going to be available for sharing amongst all the people,' he says.

Such a system needs to be able to adjust autonomously to changes in capacity; if there is a long period of drought, the entitlements system must be able to adapt without a complex renegotiation. But it's not always as simple as just cutting back all users' entitlements in proportion to the reduction in water available, says Young. For example, if the amount of water drops by two-thirds, reducing every individual's entitlement by two-thirds, while intuitively obvious, is not necessarily the most sensible course. 'You're better off putting in place a mechanism where some people take all of what's available and some people take none.'

A water-entitlement system must also look after the water needs not just of people but of the environment as well. Young says we are becoming more sophisticated in the way we talk about 'environmental' water, in that we're appreciating the difference between water for the environment and water for conveyance. This base flow – what Young calls *maintenance water* – is the engine of the river. For example, in the case of the Murray-Darling river system in Australia, maintenance water plays a vital role in keeping the river open to the sea to eject waste. But in the past maintenance water was allocated to other uses.

Another of the many challenges associated with applying property rights to water resources is deciding where to measure the resource. One common way is to allocate water into dams and then to share it from the dam. When there are no dams, another method is to apply the sharing rules to the volume of water flowing over a fixed point in a river, such as a weir. 'They'll have rules that say when the flow is greater than *x*, then you have an entitlement to pump *y*,' says Young.

Water-entitlement systems are further complicated by the relationship between ground and surface water. The more groundwater that is consumed, the less water enters the river, so in times of drought something has to give between the use of groundwater and the use of river water. Despite all these complexities, there has been a clear acknowledgement of the importance of getting some sort of system in place to manage dwindling water resources. 'I think people are getting the fact that it's necessary,' says Young, who, as one of the few people with his kind of expertise in water-resource management, is being 'swept off his feet' with requests to speak all around the world. While debate will be understandably fierce between the many people whose lives and livelihoods depend on water resources, Young says it's a step in the right direction. 'It's always better to be approximately right than comprehensively wrong.'

By establishing property rights over ecosystem services such as water, we may avoid the worst of Garrett Hardin's scenario. And as we define who owns and benefits from ecosystem services, we will start to see this value incorporated into the world economy.

Doubling the world economy

In 2005 the World Bank Millennium Ecosystem Assessment released a report on 'Ecosystems and Human Well-Being', which found that over the past fifty years humans have changed the ecosystems around us more rapidly and extensively than at any other time in human history. We have benefitted enormously from this state of affairs, but at a substantial cost. Around sixty per cent of the ecosystem services examined by the report were being used in an unsustainable fashion and were suffering degradation, mainly because of our increasing quest for supplies such as food. Another way of putting this is that we have been enjoying the benefits of these ecosystem services without really giving them a value in the world economy and this situation can't continue indefinitely.

So how much are these services worth? A group of experts has attempted to put a figure on the total value of all the ecosystem services provided by the biosphere.[24] They included services such as soil formation, waste treatment, pollination, gas and climate regulation, genetic resources and food production in a range of biomes, such as coastal areas, tropical forests and wetlands. Their research and calculations yielded the staggering figure of US$33 trillion, with a possible range from US$16 trillion to US$54 trillion. Approximately sixty-three per cent of this value was contributed by coastal systems, with the rest coming largely from terrestrial forests and wetlands. To put this value into context, at the time this paper was published (1997), US$33 trillion was 1.8 times greater than the *combined* GDP of all the countries in the world.

Instead of taking ecosystem services for granted, we are slowly coming to appreciate their importance for our survival. We are finally creating institutions that price and trade these resources, at the same time as we are creating incentives to protect and steward them. One place where this has been particularly successful is Costa Rica.

This Central American republic is a haven of biodiversity, home to an estimated 500,000 species, accounting for around five to seven per cent of the total number of species on Earth.[25] With such a range of biodiversity, Costa Rica has proven a valuable hunting ground for bioprospectors – those seeking commercially valuable chemical and genetic resources from nature, often assisted by traditional medical knowledge of the native people of an area. In the past, bioprospectors – or biopirates, as they are less fondly described – have made great use of this wealth without offering any compensation, or a share in the profits, to the guardians of that knowledge or the nations where it originated. But all that changed in 1991 with a landmark agreement between pharmaceutical giant Merck and Costa Rica's National Biodiversity Institute (INBio).

INBio was established in 1989 as a non-government, non-profit research and biodiversity management centre. Its goal is to 'support efforts to gather knowledge on the country's biological diversity and promote its sustainable use'.[26] But such a goal requires funding, and Costa Rica, while relatively well-off compared to other developing nations, has had to find the income to support INBio's aims. Thankfully, in Costa Rica, whose name translates to 'rich coast', money really does grow on trees.

According to the initial two-year agreement between INBio and Merck,[27] INBio provided Merck with insect, plant and environmental samples, collected and processed by its extensive team of staff and field personnel, many of whom were lay persons living in rural areas, and Merck had the exclusive right to examine these samples for a certain period of time. In return, Merck paid for the training of INBio staff to work at its own facilities, paid US$1 million for research, US$130,000 for lab equipment and, most importantly, agreed to pay INBio royalties on the sale of any pharmaceutical products that were discovered and developed from INBio samples.

The initial agreement has since been renewed multiple times, and time will tell whether a blockbuster drug will emerge from it. But in the meantime, INBio and Costa Rica have benefitted enormously from the injection of funds, which has helped them further INBio's conservation aims. It's an ideal scenario, in which industry draws profits from an ecosystem service but those profits are intimately linked with the preservation of that same service.

What will happen as we begin to include the inherent value of the water, soil, biodiversity and many other services in the world economy? At the very least, we'll unlock trillions of dollars of potential value to be exploited in resource efficiency. Couple this with the incorporation of environmental externalities, and you may have the greatest shakeup of the global financial system since the Industrial Revolution. Ultimately, however, profit can only go so far as a motivator. What many are coming to realise is that the costs of action on environmental issues are nothing compared to the costs of inaction.

The activities of businesses and governments can have both positive and negative community and environmental benefits, and it's in this area that the final institutional shift is occurring. Not content to just wait for governments to bring in carbon-trading systems, consumers are starting to take matters into their own hands.

Green is the new black

Driving through the pleasant, sleepy streets of Bundanoon, in the highlands south of Sydney, you'd be forgiven for thinking that not much would stir the political pot in such a laidback corner of the world. But looks can be deceiving. Bundanoon has proven itself to be a hotbed of environmental activism.

On 8 July 2009, 356 of the town's 2000 residents crammed into the local town hall for a historic vote on whether to ban bottled drinking water from their town. Locals had first considered the idea of a ban when a bottled-water company applied to extract water from a bore nearby. There was fierce opposition to the idea, for a myriad of environmental reasons. One local businessman decided to take the issue a step further and suggest banning bottled water altogether. The vote was a resounding 'yes'.

The residents of Bundanoon are not alone in taking environmental matters into their own hands. In 2003 Coles Bay in Tasmania was the first Australian town to ban plastic bags, and in 2007 the English town of Modbury in south Devon became the first town in Europe to do the same. In each case, residents were wholeheartedly in favour of the ban; by some accounts, they are quite zealous in ensuring that their towns are not sullied by unwanted plastic, whether bottle or bag.

These communities are at the vanguard of an enormous cultural shift, which is placing environmentalism front and centre in consumers' minds. 'Green' is becoming all the rage, and it has profound implications for anyone wanting to do business in this new paradigm. As communities take steps towards reducing their environmental impact, they are expecting the same – or an even greater – commitment by the corporate world. And they are using their consumer power to ensure it happens.

Every year since 1999, the research company Globescan has conducted an international survey of attitudes towards corporate social responsibility. Among other things, it examines changing attitudes towards the environment and corporations' environmental responsibilities among consumers, employees and business stakeholders. According to Globescan's 2007 report, which surveyed more than 25,000 people in twenty-five countries, environmental considerations are now a major factor in consumers' purchasing decisions.[28] Nearly one-quarter of all those surveyed – and particularly respondents from North America, Canada, Australia and Europe – said they regularly avoided a particular product or brand for environmental reasons. The survey also explored how empowered people felt to make a difference to the amount of environmental pollution created. Australians, North Americans and most Europeans considered themselves very empowered, which the report's authors interpreted as a clear sign that these consumers were happy to wield their buying power to punish companies that produced environmentally harmful products.

Interestingly, for 'green' consumers, it's not just important that they are buying products that are supposedly less harmful to the environment, but also that they are *seen* to be buying environmentally responsible products. This effect is known to evolutionary psychologists as *competitive altruism;* the idea is that to be seen to be doing good despite the personal cost adds to your status in society. The greater the personal cost of your good deeds, the greater the status.[29] This might explain why Toyota's hybrid vehicle, the Prius, has outperformed the Honda Civic hybrid: while the Honda looks just like any other Civic, the Prius has a new and distinctive look that immediately identifies it as a hybrid vehicle.

Likewise, competitive altruism can be invoked as one reason for the popularity of reusable shopping bags in supermarkets – if the reusable bags looked the same as plastic bags, they might not have been nearly as popular.

There's nothing quite so effective as this herd mentality – as long as it's moving in the right direction. A recent study found that guests in hotels were more likely to reuse their towels instead of requesting fresh ones each day if the notice said that the majority of other hotel guests reused their towels, rather than simply appealing to their environmental conscience.[30]

The writing is on the wall for businesses that don't consider the environment in their operations. It is now very clear that being a responsible corporate citizen is not only critical to a business's success, but also to its attractiveness to investors and employees.

The changing face of business

Even Wall Street – that monument to resource-driven economic growth – is seeing the green light. In September 1999 the first of the Dow Jones Sustainability Indexes was published, tracking the performance of companies leading the field in terms of corporate sustainability. So how does the Dow Jones define sustainability? 'Corporate Sustainability is a business approach that creates long-term shareholder value by embracing opportunities and managing risks deriving from economic, environmental and social developments.'[31] In 2009 the companies that topped the Dow Jones Sustainability Indexes included German motoring giant BMW, Australian banking group ANZ, Danish healthcare company Novo Nordisk and Norwegian oil and gas company Statoil. And investors like what they see – assets linked to the Dow Jones Sustainability Index have recently topped US$8 billion.[32]

It's a sign that pragmatic investors are seeing waste emissions or consumption of ecosystem services as contingent liabilities – things that may one day have a cost. Some companies are realising in return that getting a handle on their resource consumption early may give them an edge over their competitors.

But not all companies and industries are heeding the warning signs. The CEO of WWF Australia, Greg Bourne, who in a former life was head of BP Australasia, believes that a war is currently raging in industry. 'It's a war between those who are strongly invested in both the past and present, probably with huge sunk costs, and possibly with nowhere to go at least in their imaginations ... and on the other side of the battle are those who are looking very much at the future, investing their ideas in the future and progressively investing whatever money is around at the moment in new ideas,' he says. 'There are some big winners and some big losers.'

On the losing side are companies which, Bourne says, know they will most likely be out of business in twenty years; their aim is to maintain the status quo and hang on to their business for at least eighteen of those twenty years. 'Then there are a few of them who are genuinely trying to morph into something different and find a way forward.' Bourne's former company, BP – which originally stood for 'British Petroleum' – now markets itself as 'Beyond Petroleum', thanks to a massive rebranding and repositioning effort in 2000.

'The "beyond petroleum" strapline was primarily because they'd been looking thirty to forty years ahead to a climate-constrained future and realised there was going to be a time when people would be thinking beyond petroleum,' Bourne says. 'The idea was to get ahead of the curve, invest in new markets, renewable energies and so on.' The effort has received both bouquets and brickbats from green groups, with some pointing out that BP spent more on its rebranding than it spent on renewable energy at the time, but Bourne says change requires care and persistence. 'The pace at which you can do that is still limited – you can't turn business around overnight.'

There's no doubt that consumers have high expectations of industry when it comes to environmental responsibility and performance. Globescan's research shows a significant and steady increase over the past decade in people's expectations of how industry should perform in the social-responsibility stakes. For example, in 2007, in all countries surveyed except Turkey and India, the majority of respondents believed companies should be totally responsible for ensuring that their products and services do not harm the environment. Unfortunately, though, consumers are also somewhat cynical, and the perceived performance of industry on social-responsibility issues has declined. The trend seems to be that industry is not stepping up to the plate quite as much as customers, stakeholders and employees would like.

Put these trends together with the increasing cost of natural resources, and it's clear that there is a big incentive for businesses to change the way they account for their operations.

Sustainability reporting is becoming standard, and mechanisms such as the 'triple bottom line', by which companies report their environmental and social as well as financial outcomes, are the heralds of future accounting practices.

A full-value future

So what does a world in which the true price of everything is factored in look like? Probably very different from ours today. Things that were previously intangible will be added to the prices of the products that we consume, reflecting their environmental or social impact.

In a future world, consumers will be able to see the true cost of whatever they buy. Imagine reaching the checkout with your basket of goods, which, when scanned, shows not only the dollar price of your items, but also the total greenhouse gases emitted, litres of water consumed, pollutants released and number of jobs created by their production.

Our transition to this future world will not be easy – institutional change rarely is. Business models for firms that currently fail to take waste into account will have to change completely. The price of goods will go up to reflect their true costs, but the costs of the externalities borne by us all will go down. Unsustainable businesses and industries will go under, but a whole new generation of green businesses and green jobs will be created. Rather than bearing the costs indirectly through taxes and insurance, we will be closing the loop and parking those costs right back where they originated.

Will it be worth our effort? Yes, not only because it makes costs more transparent, but also because it creates an additional incentive for resource efficiency, and turbo-charges the market drivers discussed earlier. The transition will also create a massive incentive for investment in technology that will enable us to reduce or even eliminate waste, and to use our resources more efficiently. These technologies are the third pillar of the sixth wave.

Although Dorjee Sun and others like him have found a niche in helping industry to cope with the financial burden of their externalities, it may be that even those businesses will have a limited shelf life, as we develop technologies that get rid of the externalities altogether.

Chapter 5

Cleantech: from dotcom to wattcom

As elements go, silicon isn't that much to look at. It doesn't have the warmth of gold, the sparkle of crystalline carbon or the beautiful canary yellow of sulphur. But silicon's rather unremarkable surface belies its true value, which is concealed within its atomic structure.

Crystalline silicon consists of a simple lattice of silicon atoms – each one bound to its four neighbours. This structure makes it very difficult for electricity to flow through it, because there are no 'free' electrons that can move between atoms and transmit current. But if you throw into the mix a tiny amount of a carefully chosen impurity, such as boron or phosphorus, silicon suddenly becomes a whole lot more interesting. These impurities allow electricity to flow in a controlled fashion, making silicon the best semi-conducting material in the business. This is the property that made silicon so integral to the technological advances of the fifth wave, as it was the perfect foundation material for the computer chip.

Now it looks like silicon is going for a double act. Silicon's semiconductive properties make it the ideal material to absorb the sun's energy and convert it into electricity. After having dominated the fifth wave, silicon has become indispensable to the solar-power industry, a key element of the sixth wave's push towards a renewable, sustainable world.

Just as so many entrepreneurs built their fortunes on the back of silicon in the information and communication technology wave, a new generation of technologists are doing the same in the resource-efficiency wave. One of those is Zhengrong Shi. Shi is China's 'Sun King' – one of the richest men in the nation, worth an estimated US$2.9 billion[1] – and all on the back of clean technology. After earning his master's degree in laser physics in Shanghai, Shi travelled to Australia to do a PhD in electrical engineering at the University of New South Wales. There he met Martin Green, the executive research director of the ARC Photovoltaics Centre of Excellence, and began working in the exciting new field of solar cells. By 1995 he was director of a university spin-off company that developed next-generation solar cells, but Shi's sights were set on China. He returned to his homeland and set up the solar-cell company Suntech Power, then rode the wave to success.

Suntech Power is now worth an estimated US$6 billion, and was the first private Chinese company to list on the New York Stock Exchange.[2] Shi's extraordinary success has come about not only because of his incredible drive and business ability, but also because of his timing. Whether by luck or by design, Shi has found himself at the forefront of a new frontier of technological development.

The last two chapters showed how two trends – towards resource efficiency and green institutions – are shaping the world in which we live. Together, these two factors, along with markets and institutions, give rise to the third great pillar of the next wave of innovation: a massive incentive to create new technologies that allow us to measure, manage and better utilise our scarce resources.

The sixth wave is about far more than simply finding new ways to generate energy, although there's no doubt that's a major part. It's about minimising inputs, such as fuel or water, while maximising all the right outputs, such as energy, food, products or services, and minimising or eliminating all the wrong outputs, namely waste. The sixth wave starts with new technologies to manage energy, water and waste and ends with whole new business models and ways of creating value from sharing, reusing and better managing resources. Together, these technologies are known as *clean technologies,* or cleantech for short.

Just as the fifth wave was built on a suite of technologies – from microprocessors to routers to software – the sixth wave will feature a dazzling array of innovations, some of which are already emerging. The fifth wave was all about information and communication technology. In the sixth wave, efficiency is the name of the game.

Improving efficiency

Resource-efficient technologies are materials, equipment, software and systems that increase performance, productivity or efficiency while decreasing costs, resource inputs, energy consumption, waste or pollution. In simple terms, they're about getting more (and cleaner) bang for your buck.

Improving energy efficiency might not sound like the sexiest strategy to save the planet, but consider this: a recent report by McKinsey Global Energy and Materials concluded that if a comprehensive program to tackle energy efficiency were rolled out across the United States, it could deliver energy savings worth more than US$1.2 trillion, reduce energy consumption by twenty-three per cent by 2020, and deliver greenhouse gas savings of around 1.1 gigatons each year.[3]

With potential savings of that size, there is considerable interest in technologies that might deliver these efficiency improvements. There are so many different technologies under development at the moment that it's hard to cover them all, but a quick tour of the cleantech world gives us an impression of the diversity of technologies being developed. They all have one thing in common: they somehow increase the efficiency with which we use our resources.

Given that so many of our current environmental problems stem from our existing energy technologies, our tour logically starts in the energy sector. There are a large number of technologies now under development that make existing energy-generation technologies, including the coal-fired power station, more efficient. The goal is to generate more energy from the same, or a smaller, amount of resources while also giving off fewer emissions. These technologies range from developments that increase the thermal efficiency of power plants – the efficiency with which they convert fuel to heat – to projects to extract extra energy from things that were previously released into the environment, such as steam. Other technologies capture waste products, from toxins to carbon dioxide, from the energy generation process to prevent them from harming the environment.

Once you generate the energy, you want to make sure that you use every last drop of it. Unfortunately, that's not always what happens. In 2001 in the United States, around ten per cent of mains electricity was lost during transmission and distribution, as the nation's ageing power networks struggled to keep up with increasing load and demand.[4] Pressure on the power grid will only increase with time, so, in response, new technologies are emerging. One is the development of high-voltage direct-current transmission systems that dramatically improve the efficiency and security of electricity transmission.

As we continue further along the energy chain, we find that technologies to improve energy efficiency are being embedded in everyday appliances. For example, 'smart' fridges and washing machines are being developed that can choose when to switch themselves on and off, according to the amount of demand on the electricity grid. And technologies are being developed to allow the grid itself to become more intelligent, working with your appliances and energy-providers to control both the supply of, and demand for, electricity.

The potential for energy savings with these smart technologies is enormous. Already, in the United States, delivered energy use per household is declining at an average annual rate of 0.6 per cent, despite the increased size of houses and increased use of electrical appliances.[5] The US Department of Energy has forecast that by 2030, if consumers choose the most energy-efficient appliances available, residential energy consumption could be cut by as much as twenty-nine per cent.

But it's not just the energy domain where resource efficiency is the name of the technological game. Water is another resource that is not only diminishing but also extremely energy-intensive to collect, purify, transport and process, and so it offers an enormous opportunity for efficiency gains. There was a time, in the not-too-distant past, when the height of sophistication in water efficiency was nothing more than a half-brick inside the cistern of a toilet. It was simple, if a little crude, but highly effective, as it significantly decreased the water consumed with each flush.

Now, technologies that reduce water consumption at home by better control of demand, or by aeration of water, are common. We have half-flush toilets and water-aerating showerheads, and most water-using appliances come with a water-efficiency rating, similar to the energy-efficiency ratings on whitegoods. Technologies are emerging that may dramatically reduce water use even further, such as the 'waterless' washing machine, or the waterless urinal, which does away with it altogether.

Moving up the pipeline, we come to technologies that improve how we capture, store, purify and recycle water. There are a range of new ways to source water, with cleaner catchments, desalination plants and storm-water capture. Once we have captured the water, underground aquifers are being used to store and purify it, and a protective coating on the surface of dams is improving how we store water by dramatically reducing evaporation.

Reducing water use is just one part of the water-efficiency equation. An equally important element is finding more efficient ways to capture and reuse waste water and an entire industry is springing up around technologies to recycle this precious resource. The same is happening with other forms of waste, as tools emerge to process it, extract valuable materials, nutrients and energy from it, reuse it or recycle it. Technologies already exist to completely reuse or recycle 100 per cent of everyday products such as plastics, car tyres and ink and toner cartridges and ever more are being developed to 'close the loop'.

Landfills have suddenly become extremely valuable as the price of raw materials rise, so technologies are being developed that enable these once-neglected piles of waste to be mined for their treasures and treated to extract biomass and biogas that can be converted into energy. Not even the waste we flush down the toilet is exempt from this new-found gold rush – the power of earthworms is being used to convert our sewage sludge into valuable fertiliser.

With all of these clean technologies emerging, the opportunities for improving our resource efficiency are seemingly limitless; after all, very few processes are even close to 100 per cent efficient. But to make improvements in efficiency, first you have to find, measure and monitor the inefficiencies. And to do this, you need technology.

Keeping track of the world's resources

There is an old saying: if you can't measure it, you can't manage it. This is particularly pertinent for sixth-wave thinking. To manage our resources efficiently, we need to be able to measure and map them, and understand how they are used.

At three am, very little stirs in most households. The lights are off, the television and stereo are quiet, and computer screens are dark. Almost everything, whether man or machine, is asleep. It's the witching hour, the time when, as European folklore tells us, the creatures of the supernatural – demons, ghosts and witches – are at their most powerful. The storytellers might have been onto something, because in the modern world this is the hour of vampire power.

You would think that with your appliances in a state of hibernation, your power usage would be practically zero. Unfortunately, that's rarely the case, thanks to so-called 'electricity vampires'. These are appliances that draw a steady stream of power even when their owners think they are switched off. Typical electricity vampires include devices with remote controls, appliances with permanently illuminated digital displays, and power adaptors. In the average family home, there can be as many as twenty devices drawing power in standby mode. It might only be a couple of watts here or there but, given the number of appliances and the number of homes containing them, it all adds up. Around five to ten per cent of the electricity consumed in most homes is chewed up by appliances in standby mode, and the International Energy Agency estimates that standby power is responsible for roughly one per cent of global carbon dioxide emissions.[6]

The perils of standby mode for electrical appliances have been highly publicised in recent years, thanks to campaigns such as the 1-Watt Plan – a push by the International Energy Agency to reduce standby power to no more than one watt per device. Consumers have also been encouraged to switch devices off rather than leave them in standby mode, but electricity vampires can be cunning and not always easy to identify.

This is where Google's PowerMeter will come in handy. Google is developing a program which can securely link to the electricity meter in your home and deliver you an hour-by-hour account of your home's electricity use. All you need is an electricity meter sophisticated enough to provide Google with the data.

While Google PowerMeter is still in the testing phase, the Google employees involved in testing the system have made some interesting and valuable discoveries. One employee noticed regular large spikes in his electricity usage that didn't correspond to any obvious appliance activity within his apartment. He soon discovered that the washing machine and dryer in the building's communal laundry were linked to his meter. Another employee put his stereo, DVD player and VCR onto switchable powerboards, put his computer into standby mode when he wasn't using it and cleaned the coils of his refrigerator; he managed to save around US$150 a year in electricity bills.

Measuring doesn't always just mean quantifying a resource; it can be about tracking a resource as well. Global positioning systems (GPS) have become almost indispensable to modern life – GPS receivers are now built into just about every mobile phone, they are found in an ever-increasing number of cars, and are even being mooted as a way to keep track of errant dogs and children. But they also offer a great opportunity to improve efficiency, as one servicer of air conditioners in the United States found when it installed GPS navigation devices in its fleet of vehicles.

Before installing the system, the company would create a schedule for drivers at the start of the day. If service calls came in during the day, jobs would be assigned to drivers randomly. The company had no way of knowing where its drivers were at any particular time, so a service job on one side of town might be allocated to a driver on the other. This meant a lot of fuel and time was wasted as drivers spent hours driving all over town. It was a cost not only for the company but also for the environment, given the amount of fuel consumed and emissions produced. Once the company installed the GPS tracking and navigation devices, the dispatcher was able to see where all the drivers were, and so could allocate a job to the nearest driver. The savings were immediate. Petrol consumption per vehicle dropped by an average of around five litres per day, which after just six months of using the system translated into a cost saving of US$21,000 for petrol and more than US$100,000 in saved labour.[7]

All these technologies have something in common – they look for inefficiencies in the way resources are being used or allocated, and find a way to reduce or to make something of any waste. The end result is always a saving of resources. But another area of technological development is thinking completely outside the square, finding entirely new and far more sustainable resources to fuel our modern world.

Tapping in to more resources

Gyms are strange places. On any given day there will be large numbers of lycra-clad people running, walking, climbing invisible stairs, pumping and generally having a good sweat while staring blankly into space and going absolutely nowhere. If you didn't know better, it would seem like an exercise in futility. People pound away, expending valuable energy – both their own and that produced by the burning of fossil fuels – with little to show for it but red faces.

One gym has realised the value of capturing that otherwise wasted energy and putting it to good use. The Green Microgym in Portland, Oregon, has modified some of its exercise bikes so that as customers work out, they are actually generating electricity, which not only powers some of the equipment in the gym, such as the televisions, but also feeds back into the grid. You can even try this at home with the Pedal-A-Watt – a home exercise bike that can generate as much as 320 watts from a strenuous workout.

It's a small-scale example of how sixth-wave innovation is enabling us to think outside the coalmine to generate energy. One of the biggest, and potentially most lucrative, clusters of clean technologies relates to new resources for energy generation, especially resources that are abundant and as close to infinite as possible. The Silicon Valley venture capitalists who made their fortunes in the information and communication technology wave are wasting no time in pursuing clean energy generation. One of Silicon Valley's most famous and influential figures, venture capitalist John Doerr, recently described energy as 'the mother of all markets', worth an estimated US$6 trillion per year.

It's easy to think of the natural resources that can be tapped for their energy – sun, wind, oceans, rivers, hot rocks – but harder to create the technology to do so efficiently and without causing unwanted environmental side-effects. However, with that much money on the table, the ideas are coming thick and fast.

The concept of wind power usually conjures up images of huge white wind turbines jutting out from stark landscapes, their blades turning slowly in the breeze. But as tall as these monumental devices are, they are actually too short to make the most of wind power. At the height of a typical wind turbine, around eighty metres above the ground, the average wind speed is around 4.6 metres per second. The amount of wind power available is related directly to wind speed, and at this height it's about 58 watts per square metre. But go up ten times that height, to 800 metres, and average wind speed increases to 7.2 metres per second, which can generate around 205 watts per square metre.[8] That's a significant increase in potential energy, with the added advantage that the wind is more dependable at higher altitudes. But building a wind turbine about the same height as the world's tallest building would be an almost impossible feat of engineering. One wind power company, however, has come up with an altogether freer solution.

According to Italian company Kite Gen, the most efficient parts of a wind turbine are the wingtips, as they reach the greatest speed; ninety per cent of the power generated by a typical three-blade turbine comes from the outer forty per cent of the blade. So why not do away with the rest of the turbine altogether?

Kite Gen's wind power model consists of a semi-rigid, automatically piloted kite, designed to capture the maximum amount of wind. The kite is tethered on either side by two high-tensile cables. On the ground, these cables are wound around two drums, which are linked to generators. The system works a bit like a yo-yo. As the kite plays out, the cables unspool, spinning the drums, which drive the generators to create electricity. When the kite reaches its maximum height, a small engine reels it back in, only using up around twelve per cent of the generated power to do so. Then the whole process repeats itself.

Kite Gen's simulations suggest that a single kite of 100 square metres in size, on a 300-metre line, could generate an average of 620 kilowatts. According to the company's calculations, a kite power plant of sixty or seventy kites of about 500 square metres in size, attached to a carousel with a three-kilometre diameter and flying 800 metres above the ground, could generate an average of nine terawatt hours per year, at an estimated cost of US$15 to $20 per megawatt hour. By comparison, to generate the same amount of energy using wind turbines you would need 2250 turbines, which would cost US$140 to $160 per megawatt hour. To produce the same amount of energy using fossil fuels would cost US$80 to $90 per megawatt hour.

Kite power has the added advantage of requiring much less land space. A wind farm like the one described above would occupy just eight square kilometres of land, compared to 300 square kilometres for the turbine wind farm. Kite farms could also be set up in areas that are unsuitable or inaccessible for land-based wind farms, because their height makes them less vulnerable to the vagaries of wind close to the ground. You could even set up a kite power farm above land already occupied by other structures, such as in the nofly zones above nuclear power plants.

While kite power is looking to exploit the air, wave power is exploiting the water. Five kilometres off the coastline of the Portuguese parish of Aguçadoura lies the world's first commercial wave farm. From the air, it looks like relatives of the Loch Ness monster are making their way to shore. Three 150-metre red tubes lie along the surface of the water, bobbing slightly in the swell, but these deceptively simple devices are harnessing the energy of every wave that laps along them.

These tubes are Pelamis Wave Energy Converters – the creation of Scottish firm Pelamis. Each is a semi-submerged, articulated structure of carbon steel, made up of a series of cylindrical sections connected by hinged joints. Inside those hinged joints are hydraulic rams designed to resist the up and down movement of the cylinders. As a wave passes the length of a tube, the sections bob up and down relative to each other. This motion pushes the hydraulic rams, driving hydraulic fluid at high pressure, which in turn drives the generators that produce the electricity. This one farm of just three devices can generate 2.25 megawatts – enough to meet the typical electricity needs of more than 1500 houses – and the company is planning to expand the wave farm by a further twenty-five devices, making it capable of generating up to twenty-one megawatts of power.

Wind and waves are almost limitless resources – as long as there is an atmosphere around the planet, created by the heat of the sun, and oceans covering its surface, controlled by the gravitational pull of the moon, there will be plenty of both. But even after the atmosphere and the oceans have been boiled or blasted away by cosmic forces, the enormous engine at the heart of our solar system will labour on.

Solar power is perhaps the most promising of all renewable energy resources. The Sun burned brightly long before life even existed on our planet. It will almost certainly preside dispassionately over the demise of humankind, and by the time its light finally dies our very planet will no longer exist, having being vaporised in the Sun's death throes.

Professor Andrew Blakers believes solar power has it all. 'If you are looking for the characteristics of the energy source you need, it's got to be huge scale, otherwise it can only make a small contribution, and that rules out all energy sources except coal, fast-breeding fission, fusion perhaps in fifty years, and solar photovoltaic and solar thermal,' says Blakers, who is the director of the Centre for Sustainable Energy Systems at the Australian National University. Solar power has the added advantage of having no military applications, so there will never be the risk of wars fought over and with it. And it is, for all intents and purposes, unlimited. 'We will never ever possibly, even in the remotest future, run out of the raw materials for solar, because that's what the world's made of – silicon,' he says.

That is not to say that the kind of pure silicon needed to produce solar cells can be found on every street corner, so in the spirit of resource efficiency Blakers and his colleagues have developed a type of solar cell that uses up to ninety per cent less silicon than existing cells. These Sliver cells are also extremely thin, making them very flexible and almost transparent, opening up a whole raft of possible uses. For example, Sliver cells could replace the glass in windows, or be used as cladding on buildings or to build solar-powered aircraft. Other companies are looking at doing away with silicon altogether, with organic solar cells.

The possibilities for solar power are limited only by the imagination of the scientists working on them. Blakers and his colleagues are also developing an all-in-one solar trough concentrator, which is designed to supply all the power, heating and cooling needs of a building.[9] The trough is composed of sun-tracking mirrors, about 2.5 square metres in size, which reflect light onto a receiver lined with solar panels. The sun's energy is concentrated so that the cells are hit with approximately twenty-five times the normal solar concentration. They convert about one-fifth of this energy into electricity, while the rest of the energy is used to heat water, for space heating and for cooling.

Using similar principles but on a far bigger scale, US company eSolar recently unveiled its Sierra SunTower. Its 12,000 mirrors reflect the sun's rays onto receivers mounted on the tower, which use the heat to boil water and create superheated steam. This steam then powers a turbine, which generates electricity. A single module covers ten acres and can produce enough power to supply 2000 households with electricity.

But even that is small fry compared to the scale of the Desertec project.[10] This ambitious scheme, an initiative of the Club of Rome, is backed by some of the world's biggest blue-chip companies. The consortium plans to build enough concentrating solar thermal power plants in the deserts of North Africa and the Middle East to meet all the electricity needs of those regions and of Europe. The idea behind Desertec is that in the space of just six hours, the world's deserts receive more energy than all the people in the world consume in a year. More than ninety per cent of the world's population live within 3000 kilometres of a desert. Solar collectors covering just three-thousandths of the world's deserts – equating to around 90,000 square kilometres – could supply the entire planet with all the renewable, CO_2 free energy it needs. According to the Desertec consortium, all the technology needed to pull off a project of this size already exists and is in widespread use. With the political will, they argue, their global energy vision could be realised in less than thirty years.

Not all resources have to be related to energy, although they are very much front-of-mind these days. As we have already seen, human resources are being tapped into in new and exciting ways. Until recently, the talent pool at a company's disposal was almost always contained within the four walls of a building. More recently, there has been outsourcing of some less-essential tasks to local small companies, and maybe a dabble in some contracting. But now there's night-hawking.

As radiologists in the United States have some well-earned sleep, trained and accredited radiologists in India are examining the day's medical images sent from America and compiling medical reports. When the US specialists get to work the next morning, they will have those reports at their fingertips and will be ready to start a new day of consulting. This is the practice of night-hawking, which is also known as teleradiology. It's another step along the outsourcing route, which has already seen so many services contracted to overseas agencies whose employees are able to perform tasks more cheaply, but also while those who pay them are asleep.

All of these technologies tap into resources that already exist and make them accessible to the market. But there is another group of technologies emerging as part of the sixth wave, which take resource efficiency to the ultimate level – consuming no resources at all.

Closing the loop

E-waste has received a lot of attention in recent years. Our insatiable hunger for the latest electronic gadget or hardware upgrade is building an ever-increasing mountain of electronic waste. This mountain contains some extremely valuable resources, such as lead, copper and gold, but it is also contaminated with substances that pose a threat both to human health and to the environment, such as dioxins, polychlorinated biphenyls (PCBs), cadmium, chromium, radioactive isotopes and mercury. One of the biggest challenges with recycling e-waste is how to get to the good stuff while avoiding the bad stuff.

Close the Loop is an Australian company that has found a way to recycle 100 per cent of a particular type of e-waste: the ink and toner cartridges used in printers, photocopiers and fax machines. These are particularly problematic to recycle because the cartridges are often very complex, made with many different types of plastic and metal, and contain leftover toner particles, which are not only extremely fine but also potentially explosive in certain conditions.

To begin with, nearly half of all the cartridges that Close the Loop receives are actually returned to their original manufacturers, who are able to replace their worn parts and reuse them. But for those that can't be returned, Close the Loop's solution is its patented Green Machine, which cracks open and shreds the cartridges, reclaims any toner via an extraction system, then separates out the metal and plastic. The metal – usually aluminium and steel – is sent to metal recyclers and, where possible, the plastics are sent back to the original equipment manufacturers for reuse in new cartridges.

What's left behind is a mix of plastics that Close the Loop's Duncan Freemantle describes as a 'contaminated plastic stream'. 'If we run a batch where we have a significant amount of mixed plastic ... you do generate a contaminated plastic stream, or what's considered contaminated in the commodities market. You've got no place to go so, being a zero waste to landfill company, we had to come up with solution,' says Freemantle, the manager of Close the Loop's technology division for the Australasia region. The solution the company came up with was eWood.

'eWood allows mixed contaminated plastic to be extruded into a useful product, like for landscaping,' Freemantle says. eWood looks and feels similar to wood products, and can be worked and shaped just as easily. It can be used in many of the same applications as wood, such as fences, outdoor furniture and garden hedging, but it's made of 100 per cent recycled cartridge plastic. It's also free of the contaminants that plague e-waste, especially brominated flame-retardant chemicals. 'We have a specialised extrusion process ... that deals with brominated flame retardants – we can extract them, ensuring that the end product doesn't have it in it,' Freemantle says.

What Close the Loop has succeeded in doing is exactly that – closing the loop, so that nothing is wasted and every part of a product goes back into the system and is reused in a productive fashion. The company is even working with cartridge manufacturers to help them design cartridges that are easier to separate into their component parts and reuse or recycle.

Companies are also trying to close the loop for oil-based resources. Molectra Australia specialises in recycling car and truck tyres, which are another major environmental hazard. An estimated 18 million waste tyres are generated in Australia alone each year; more than half of them go into landfill, and thirteen per cent are dumped illegally. They pose a threat to human health, as burning stockpiles of tyres can release toxic gases into the surrounding area. It also represents a huge waste of a very precious resource: oil.

Molectra has developed a means to recycle 100 per cent of the rubber in tyres, not only creating new products from the rubber itself but also recovering oil from the rubber feedstock, which can then be used to power the whole process with energy to spare. As Molectra starts to partner with the tyre companies themselves, all the resources in the process will be recycled into new products, which in turn can then be recycled into others.

Close the Loop and Molectra have made resource efficiency a profitable venture, and they're not the only ones. Where there is profit, there are investors. Fuelling all of this technology development are huge investments in all sectors of the economy, from government to big business and venture capital.

The new silicon valley

To be truly successful, a good venture capitalist has to be able to tell which way the innovation wind is going to blow before it even changes direction. The venture capital markets, therefore, are often where much of the initial activity of innovation waves begins. Right now, the wind is definitely blowing in cleantech's direction, and many of the world's leading venture capitalists are investing in it heavily.

In 2008, Silicon Valley investment in cleantech was valued at almost US$1.9 billion. While the global financial crisis has led to dramatic downturns in so many other industries, cleantech investment by Silicon Valley increased by ninety-four per cent from 2007 to 2009, while employment in the sector also increased by twenty-three per cent since 2005, according to the 2009 Silicon Valley Index. This is despite total venture capital investment in Silicon Valley dropping nearly eight per cent from 2007 to 2008.[11]

And this is being repeated on a global scale. Venture capital investments in cleantech increased ten-fold from 2002 to 2008 to a whopping US$8.4 billion. While investment in cleantech dropped by forty per cent in 2009 as a response to the global financial crisis, other sectors dropped even further, such that cleantech now draws the most venture investment of any sector in the United States, overtaking software and biotechnology.[12] The outlook for the sector sees this lead being cemented – according to the US National Venture Capital Association cleantech is the industry which has the most optimism for future growth, with fifty-four per cent of venture capitalists predicting increased investment in 2010.[13]

Rich Stromback is one of the Silicon Valley entrepreneurs throwing his financial weight behind cleantech. A former professional hockey player, he made his fortune in the fifth wave as the founder and CEO of the Web Group, an IT staffing and search company. But by 2003 momentum in the information technology boom was waning, and Stromback's attention was captured by the nascent cleantech industry. At the time, the big guns of Silicon Valley were yet to take up the charge, says Stromback, but the many opportunities and challenges of the cleantech industry caught his imagination.

When looking around for what to invest in he discovered the world of environmentally friendly industrial coatings. Why coatings? Because just about every manufactured product has some kind of coating on it, to make it stronger, tougher, shinier, scratch-resistant, rust-proof and so on. Unfortunately, these coatings are often hazardous not only to human health but also to the environment, says Stromback, as they release volatile organic compounds and hazardous air particles and are extremely energy-intensive. He acquired Ecology Coatings, a research and development company in the field. Ecology Coatings has developed an entirely new class of coatings based on nanotechnology that are solvent-free and curable by ultraviolet light.

But Stromback is also investing in the broader cleantech industry through his private equity firm Ecology Ventures, which invests in early-stage clean-technology companies, particularly those involved in energy efficiency. 'I get most excited about conservation technologies, especially energy-conservation technologies, because those have immediate impact but are often overlooked,' he says. 'There's all this money piled into alternative energies, but what about technologies that reduce the number of power plants we need?'

Stromback and his peers are not the only ones thinking like this. A recent report by the McKinsey Global Institute suggested that an annual US$170 billion investment in energy-efficiency technologies – such as those that improve the efficiency of lighting and air conditioning – could halve the projected growth in global energy demand every year from now until 2020, and make a major contribution toward reducing greenhouse gas emissions.[14] So it's no surprise that it's not just venture capitalists who are scrambling to invest in clean technology – the investment industry, governments and big business are all suddenly seeing green.

Big business

The story of American company General Electric reads like a history of twentieth-century innovation. As one of America's oldest companies, General Electric has weathered numerous waves of innovation; it has not only survived them but has flourished, expanding from its original beginnings in a backyard barn to become a multi-national conglomerate that deals in everything from aeroplanes to finance. In 2009 Forbes ranked General Electric as the world's largest company.[15]

One of the keys to General Electric's longevity is the way it has been able to reinvent itself time and time again, to survive the tumultuous and often destructive transitions from one wave of innovation to the next. And now it's doing it again. In 2008 General Electric invested US$1.4 billion in clean-technology research and development, as part of its 'ecomagination' pledge to invest a total of $1.5 billion by 2010. Why? Because, as CEO Jeff Immelt has emphasised, there is serious money to be made in environmental technology.[16]

Other large businesses are starting to invest in clean technologies for the simple fact that they see it as part of their future competitive advantage. Take the telecommunications industry, one of the champions of the fifth wave. Energy bills for some telecommunications companies now make up twenty to thirty per cent of their operating costs. This number is continuing to rise, to the extent that telcos are becoming power companies in their own right.

No one knows this better than one of the larger information-based companies in the world. Google's RE<C initiative, launched in late 2007, aims to produce one gigawatt of renewable energy that is cheaper than coal. The company plans to invest tens of millions of dollars towards this goal and has already started assembling a team of engineering and energy experts to lead research and development into renewable energies, starting with solar thermal energy.[17] Like General Electric, Google is not doing it simply out of concern for the planet – the

company sees the dollar signs at the end of the process, and is planning to invest even more significantly in any renewable energy projects that take wing.

Just as big business is predicting the bottom-line benefits of investing in clean technology, so too are the world's investors.

Investing in the future

Permaculture and the financial industry are two concepts that seem to be polar opposites – one concerns agriculture that is based on natural ecosystems, and the other is about money and profit, much of which is made through finding and consuming resources in those same ecosystems. But in the early 1980s in Australia, a group of people with an interest in permaculture decided to bring the two disparate forces together. They created an investment portfolio that didn't involve pillaging natural resources: Australian Ethical Investment.

The original motivation was simply for investors to know exactly what their money was doing, according to James Thier, an executive director of the company. 'In permaculture, nutrients circulate in a circular, self-sustaining manner without exogenous inputs like fertiliser and pesticides – it's a closed-loop system,' he says. 'And they said, "Look, if nutrients circulate in a circular, self-sustaining manner, why not our money?"'

With this in mind, the group set up an investment fund based on an ethical charter that has not changed in more than two decades.[18] It seeks out investments that support, among other things, the amelioration of wasteful or polluting practices, the development of sustainable land use and food production, the preservation of endangered ecosystems, and the efficient use of human waste. At the same time, the fund avoids investing in things that unnecessarily pollute the land, air or water, that destroy or waste non-recurring resources, or that involve goods or services that have harmful effects on living creatures and the environment they live in. It also addresses non-environmental ethical principles by, for example, supporting the development of workers' participation in ownership and control of their work organisation and avoiding sweatshop-type operations.

There are some broad themes to be found among the investments that AEI makes, says Thier. For example, the fund looks for investments that focus on water and water usage, such as water-efficiency products and water-metering systems; adaptive technologies, such as pollution-control devices; waste management and recycling technologies; efficient transport methods; and even education and software – because 'those sort of things often reduce resource usage, or at least change the mindset of people to think about better ways of doing things', Thier says. The companies AEI invests in must meet a stringent set of criteria, which is investigated very carefully, and AEI reviews its investments every year. In 2009, companies on AEI's list included geothermal energy company Petratherm, international organic-food retailer Whole Foods Market, and software company Adobe Systems.

Ethical investing is on the rise, as consumers extend their newfound environmental awareness beyond their own immediate actions and to the actions their money pays for. 'Lots of people now plant trees, they walk to work, they reduce their footprint, recycle their rubbish, but often they don't think of what their money is doing – and their money can be a very powerful tool,' Thier says.

Business is responding. Thier says companies now approach AEI and ask what they need to do to get on its investment list. 'All these issues have bottom-line implications,' says Thier. 'At the end of the day, if a company is doing the wrong thing it's going to be penalised, whether it will be fined by the EPA, whether its reputation is destroyed or diminished – all these things will affect the company, its profitability and ultimately its share value.' And thanks to the financial success of ethical investment funds, which generally perform as well as the rest of the pack if not better, there is a growing awareness that it makes good business sense to do so.

The investment industry is also showing an awareness of the broader context of its investment decisions, according to Nathan Fabian, the CEO of the Investor Group on Climate Change (IGCC). 'There is a realisation that they can't expect to earn financial returns in an economic vacuum, and ultimately the long-term success of investment activities is related to the strength or vibrancy or health of the natural environment and the societies which participate in the markets they invest in,' he says.

Hence the existence of organisations such as the IGCC, which represents the views of investors to policymakers, and also helps advise investors on climate-change risks in their portfolios, how to address them and how to take advantage of emerging opportunities.

Some sectors of the investment industry are ahead of the game, particularly in the superannuation sector. 'They're the ones who can see these trends coming together,' Fabian says. 'Fundamentally, they know that the question of sustainability between environment resources, social capability and economic returns are aligned.'

Part of the problem facing the investment industry as a whole is uncertainty about the policy frameworks surrounding issues such as carbon emissions, which is making it difficult for investors to enter the market with any degree of confidence. 'It's difficult for investors who manage other people's money to incorporate something as an investment-influencer without some legitimacy for that issue in the market,' Fabian says. However, once that framework is in place, the market is open for business. 'They can start to cost the carbon-emissions risk in the assets they already hold, they can forecast how that will play out and then they can start to compare it to alternative business models and investment opportunities.'

This brings us to the last – but certainly not the least – party with a stake in sustainability: the world's governments, whose action (or inaction) on environmental issues can make (or break) the planet.

Government investments

'When it rains, it pours'; so the saying goes. And in the renewable-energy industry in the United States, the monsoon has finally arrived, courtesy of President Barack Obama's stimulus plan. Approximately one-tenth of the US$787-billion package was allocated to environmental initiatives, including nearly US$47 billion to support clean-energy research and development, US$11 billion to upgrade the electricity grid and US$7.2 billion for environmental cleanup.[19] And it doesn't stop there. Congress has also appropriated US$7.8 billion for research and development of energy technologies by the Department of Energy.[20] This includes US$400 million for a programme called ARPA-E, the goal of which is to 'uniquely focus on high risk, high payoff concepts – technologies promising true energy transformations'.[21]

The United States isn't the only government investing in clean technologies. In 2008, the Chinese government dedicated nearly one third of its four trillion yuan (US$585 billion) fiscal stimulus package to low-carbon technologies, and has also invested US$2.9 billion in an electric car development program.[22] Canada has invested CAD$1 billion in clean technologies through its Sustainable Development Technology Canada fund,[23] and so the list goes on.

Government investment in clean technology is about more than simply money. Around the world, governments are making sweeping institutional changes that are intended to stimulate investment in these technologies. For example, the European Union is providing incentives to the industry through tax benefits to produce over twenty per cent of its energy from renewable sources by 2010, and the Australian government recently passed legislation mandating that by 2020, twenty per cent of the nation's power should come from renewable sources. These and many other government endeavours are creating a fertile environment for the clean-technology industry, and by doing so are attracting investors, who can see not just environmental but also financial benefits on the horizon.

Decarbonising the economy

So what will our world be like when all these technologies are rolled out? Our homes will be populated with smart appliances that switch themselves on and off, to ensure a minimum of waste and maximum of efficiency. Landfill will be obsolete, as every bit of waste we throw out will be carefully sorted and reclaimed. Water will be managed as carefully as electricity – every litre accounted for, tracked and treasured. The air in our cities will be a lot cleaner, with less particulate pollution but also with less noise pollution, as electric vehicles purr silently along our streets.

Most importantly, these technologies will change the very shape and basic drivers of our economy – for the first time since the Industrial Revolution economic growth will no longer be dependent on the consumption, and production, of natural resources. Success in this future world will require very different approaches from those of today...

Part II

Catching the wave

Part II

Introduction

If there were such a thing as a time machine and you could go back in time, what advice would you give yourself?

While no one can see the future, the first half of this book has looked at the next wave of innovation, and the way it comes about through massive changes in markets, institutions and technologies.

From the market perspective we are seeing massive inefficiencies in the way that we use natural resources. Match this with the declining availability of some of these resources and you get a huge, and as yet largely untapped, business opportunity.

At the same time, there is a trend towards global governments internalising things called externalities, which basically means putting a price on things that inadvertently have a harmful impact on the community. Institutions are being developed to protect and price natural resources that have never been priced before, from water to soil, trees to biodiversity. The amount of value sloshing around in the world economy is about to increase dramatically as we address the greatest market failure of all time: climate change.

Both of these trends are causing massively increasing investment in clean technologies: developments that will use resources more efficiently, and those that will allow us to keep track of these resources. New technologies will enable us to reduce our impact on the planet and ensure the sustainable production of energy. Governments, big business and upstart ventures are all competing to find the next 'killer app' in this new market.

In short, the next wave of innovation will be driven by resource efficiency, enabled through the pricing of waste and natural resources, and turbo-charged by clean technologies.

This is an enormous shift. Over the two hundred years since the Industrial Revolution, we have seen economic growth strongly coupled with the consumption of more and more resources. The more we grew, the more we consumed. If the next wave of innovation really will be tied to resource efficiency, this equation is about to be turned on its head.

In a world that is 'decarbonised' or 'deresourced', we will have decoupled resource use from economic growth. The world economy will no longer be based on the transformation of inputs into outputs. This will not happen overnight; if resource efficiency really is at the heart of Kondratiev's sixth wave, it will be a journey of some thirty to forty years.

So what does this mean for the shape of the future? How do you catch the wave? The second part of this book looks at five big ideas that will shape the sixth wave. In other words, this is what we believe we would be whispering into our own ears if we had a time machine from the future.

The first big idea – that waste equals opportunity – describes a way to look for opportunity everywhere. Waste is to the sixth wave what transaction costs were to the fifth, and the greater the waste the greater the opportunity. The future world might look very different when there is no waste at all.

This leads us to the next big idea: sell the service, not the product. If waste is the opportunity, services are the answer. Companies and nations are learning that the best way to create value without consuming resources is through services, and this concept has far-reaching implications for the world economy, the internet and our way of life as we know it.

We also see that the digital and natural worlds are about to converge – the third big idea. With more value in monitoring our natural resources, more and more devices will emerge that connect the digital to the natural. This is true of everything connected to the net (you already have your digital analogue – the mobile phone). Moreover, the whole planet and all of its natural resources will be measured and monitored to the point that it will be hard to distinguish between real and virtual.

Put these three ideas together and you get the fourth: atoms are local, bits are global. That is, the pursuit of efficiency will drive everything to do with natural resource consumption to become more local, while the international nature of the internet will make anything to do with information a truly global endeavour. Energy production will be distributed and localised, and resources will be recycled as closely as possible to the point of consumption. Some businesses already have a name for this: glocalisation.

Finally, there is a catch-all. If in doubt, look to nature. Mother Earth has had billions of years to learn the lessons of resource efficiency, and she still has a trick or two up her sleeve. If we just look closely we can see that solutions to many thorny problems already exist in the natural world, something known as biomimicry.

Those who get this already can probably put this book down now. But for those who are interested in taking a journey into the future, join us as we learn how to surf the sixth wave.

Chapter 6

Waste=opportunity

Step inside just about any car mechanic's workshop and you'll find a host of tools powered by nothing more than compressed air. The development of pneumatic tools has been a saving grace for many a mechanic – they are lighter, have a higher power-to-weight ratio than conventional tools, have fewer moving parts to wear out and are usually not only cheaper but also safer than their electric counterparts.

But they're far from perfect, as Australian mechanic Chris Bosua discovered when he tried to run a pneumatic piece of machinery using a compressor that was sadly underpowered for the task. While the idea of a tool running on air sounds vaguely environmentally friendly, pneumatic tools are actually demons of waste. Much of the compressed air that goes into powering them is simply vented through an outlet valve, not only wasting energy but also creating pollution in the form of noise, oil vapour and blown dust.

The solution to all these issues turned out to be very simple – turn the wasted air pressure back into power. The Exhausted Air Recycling System (EARS) is effectively a closed-loop system that takes the exhaust air from the pneumatic tool and routes it back into the air compressor intake via a second hose and special manifold. Bosua's invention not only doubles the capacity of an ordinary air compressor but also drastically improves its efficiency, because the recirculated air is already at a higher pressure so the compressor doesn't need to work as hard.

If used continuously, EARS can reduce the compressor's energy consumption by up to forty per cent. It also reduces the amount of noise made by pneumatic machinery – for example, the noise level of a pneumatic drill dropped from the equivalent of a lawnmower to a sewing machine. And because the air is being recycled rather than drawn into the compressor from the outside, it contains less moisture and is generally cooler, thus extending the life of the compressor, hose and machinery. It's a simple concept that represents a revolution in efficiency for pneumatic machinery and compressors. All it's doing is taking waste and turning it into something useful.

In the sixth wave, waste equals opportunity. If you can find waste in a system, the opportunities will follow, because in a world of limited resources waste is a huge source of untapped value. In the sixth wave, waste is simply another 'transaction cost'; in a world where resource efficiency is a source of competitiveness, waste is a good place to look. Anything that reduces or eliminates waste is likely to do well. This might take the form of a technology that gets rid of waste by improving the efficiency of a system, but it might also be a technology that enables us to extract value directly from waste.

Wasted

Waste is everywhere. Just about every item in your home, your office, your environment and on your body has created some form of waste, either in its production or its use, or it will do when it's disposed of.

One of the first to see the opportunity from all this waste was John White, the founder of waste-management company Global Renewables. The statistic that most astonished White was one released by the Environmental Protection Authority in the Australian state of Queensland some years ago, which stated that eighty per cent of all tradable goods end up in landfill within an average of six months. 'I realised that all of this material is incredibly energy-intensive to dig up, to make, to distribute, to warehouse, to shelve,' he says. 'It consumes an enormous amount of energy, and all that embodied energy in that material gets thrown out when it's put into landfill.'

So in 1998 White decided to do something to recover some of this lost energy. Having come from a mining and heavy-engineering background, he saw the problem as an issue of materials separation and beneficiation – the mining process by which valuable ore is separated from waste material. With the help of mineral-processing company Minproc, the UR-3R – standing for 'Urban Resource – Recycling, Reduction, Recovery' – Process was born.

UR-3R deals with the waste that's left over after just about everything else of value – such as paper, plastic, glass and metal – has been removed. What's left behind is still between forty and seventy per cent of the original waste stream and it all ends up in landfill. 'In effect, this is the greater problem because it contains all manner of materials, including a lot of organic material, which, when put in a landfill, rots and creates very toxic leachate,' says White. 'Rotting organic material in a covered landfill also creates methane gas, a big percentage of which always escapes into the atmosphere.'

To start with, the waste is sorted to remove any remaining recyclables, which are sent to their respective recyclers. Then any toxic items, such as car batteries, are removed. What's left is a heap of organic waste that is put through a three-stage process of separation, biological processing and composting. This not only breaks down the material but also helps to power the whole plant. 'We collect the methane and use it as a source of renewable electricity and export the surplus back to the grid,' says White. Organic waste also contains a large amount of water, which is collected and used as part of the UR-3R process. The end result is an organic fertiliser that can safely be used on farmland to increase soil fertility, and – thanks to the methane and water collection – the entire process is carbon-neutral. But it also prevents a considerable amount of greenhouse gas being released into the atmosphere. White says the estimate is that for every tonne of waste treated, a couple of tonnes of greenhouse gas emissions are avoided.

But UR-3R is about more than just saving the environment – it's also about turning waste into profit. Since building the first plant at Eastern Creek, near Sydney, which is capable of processing around 200,000 tonnes of waste per year, Global Renewables, in partnership with Lend Lease Corporation, has been tasked with managing the waste of an entire English county. The Lancashire waste contract is a major coup for Global Renewables and its UR-3R process. 'It's the biggest recycling contract ever awarded that I know of in the world,' White says. 'It will process up to three-quarters of a million tons of waste a year for twenty-five years, and total revenue is probably going to be US$3 billion to $5 billion over that time period.'

Not bad for a heap of rubbish.

One man's trash...

As the saying goes, one man's trash is another man's treasure. These days, in fact, one man's (or woman's) trash is society's treasure, as we come to appreciate the value of materials that until now we have cast away into landfill, incinerators or the environment without a second thought. Landfills also take up valuable land that might otherwise be used for more constructive purposes, so they have suddenly becoming very interesting from a commercial perspective.

A Dutch company, Multi-Purpose Industries (MPI), sees so much potential in landfill mining that it's putting its money where its mouth is. If the company comes across a landfill it believes has ingredients that make it a viable prospect, it proposes to 'buy' the waste for a token €1 and rehabilitate it free of charge; in exchange, it gets access to anything it recovers from the waste. The landfill owner gets a reborn site that can be used for urban development or industrial use, avoiding a costly rehabilitation process, while MPI gets access to a host of valuable resources. It's a win–win situation.

Landfill mining isn't a new field but it is now coming into its own, according to MPI's Jelle Frölich. 'Technically, it was possible a decade ago, but economically, it's possible now because of increasing pricing for raw materials and energy,' Frölich says. It's also encouraged by regulations in the Netherlands that require ongoing care of landfill sites.

MPI must first find a landfill that has the right balance of reusable materials and organics. 'Integrated waste mining is not profitable for every waste flow or dumpsite,' says Frölich. 'You need a minimum quantity, and the potential value of the waste must be high enough.' Once a suitable site is found, MPI sets up a demountable treatment facility either directly on or near the landfill, which drastically cuts transportation costs. Through an extended sorting process, workers extract all the reusable materials – oil, metals, glass, certain plastics and building materials, which make up around sixty to seventy per cent of the waste. What's left is biomass, which is converted into energy; around 21.5 megawatts can be generated from a landfill containing a total of 500,000 tonnes of waste. A small amount of toxic substances are destroyed or immobilised on-site.

The process, which is still at the pilot stage, results in nearly zero emissions, and the only material that remains at the site is sanitised soil, which can then be reused to build whatever urban project is on the cards.

Landfill mining is all about yesterday's waste, and how to extract value from what we once considered valueless. But what are we doing about today's waste?

Waste as unsellable production

You'd be forgiven for thinking the Singaporean island of Pulau Semakau was just another typical tropical island paradise. There are flourishing mangroves, a large seagrass lagoon, a thriving coral reef and a host of tropical birds. But Pulau Semakau is no ordinary tropical island – it's actually a garbage dump.

Space is at a premium on the island nation of Singapore, so rubbish disposal has been a particularly vexing problem. By 1999 the last remaining landfill on the main island of Singapore was approaching full capacity, raising the question of what to do with the 5.6 million tonnes or so of rubbish generated by Singapore's 4.4 million residents. Firstly, the government implemented a comprehensive recycling program, which takes care of around fifty-four per cent of the waste. Around forty-three per cent would be incinerated, and the remaining three per cent sent to landfill. But there was still the problem of the ash left over from the incineration.

The solution was Pulau Semakau. It is actually two islands joined together by a rock embankment, which creates an enclosure between the two. The space inside the embankment is divided into eleven bays, or cells. The cells are where Singapore's incinerated rubbish ends up. As each cell is required, the seawater is pumped out and the cell is lined with heavy clay and thick plastic to prevent anything from leaking out into the surrounding waters. The ash is poured in, and when a cell is full it is covered over with dirt and planted with grass. So far, four of the eleven cells have been filled.

But Pulau Semakau has become something far more than just a neat solution to a waste problem. When the island was first created, some mangroves had to be destroyed to make way for it. They were replanted near the site, and have now grown so much that they actually need to be trimmed back. The island's ecosystem has not only survived but is flourishing. Dolphins, otters and green turtles have been seen in the waters around the island, the coral reefs hum with marine life, and the mangroves harbour numerous species of birds, some of them rare. In an unlikely twist to the island's story, tourists now visit Pulau Semakau to sample its rich biodiversity.

Singapore found a way to turn its waste into something constructive, but it's not exactly the most profitable example. While turning waste into a biodiversity hotspot has solved some major logistical problems facing Singapore, many other small countries are managing to turn their waste into money. What many businesses and individuals are now realising is that waste isn't waste at all, but simply another form of production. At the moment, waste is usually thought of as the dead zone of profit. But the sixth wave is changing this, and industry is coming to appreciate that waste can be sold. Some businesses have even emerged that match one company's waste with another company's want.

The iridescent shark, *Pangasius hypophthalmus,* is not actually a shark but a catfish, and a tasty one too. It is one of the most popular eating fish in Vietnam, and is the main product of the Hiep Thanh Seafood processing plant in Vietnam. This plant turns whole fish into fillets, steaks and pieces, producing 120,000 kilograms of fish waste in the process every day. Until recently, the company sold this waste to the feed industry. The fish waste has a particularly high fat content – around twenty-two per cent – which makes it an ideal candidate for conversion into biodiesel. Accordingly, a European consortium, led by the VTT Technical Research Centre of Finland, is constructing a biodiesel plant next to the fish processing plant, so that the fish waste can be converted directly into fuel. Not only will the biodiesel help to power the plant and the conversion process, but some energy will also feed back into the local grid.[1]

Waste reuse is not only about generating energy. It can also be about generating an entirely new product. For example, brewing beer is a particularly wasteful process. A single litre of beer takes around twenty litres of water to produce, and the vast bulk of the grain used is discarded – a huge waste of valuable protein and nutrients. Unfortunately, the brewing process changes the grains so they are generally unsuitable for animal feed, but a brewery in Canada, Storm Brewing, has found an even more profitable use for their spent grains – growing oyster and shiitake mushrooms.[2]

But it doesn't stop there. Growing mushrooms on the spent grains actually changes the grain waste, making it digestible by livestock, so after the mushrooms have been harvested, the waste left over from that process can be fed to animals. Then the waste produced by those animals can be combined with waste water from the brewery to generate biogas and a solution rich in nutrients. The biogas can be captured and used for fuel, while the nutrient solution can be used to grow algae for fish food.[3] So, using the waste from brewing beer, we get mushrooms, animal fodder, biogas and fish food.

Then there's the waste associated with the consumption of beer – or with the consumption of any food or drink, for that matter. Human waste is plentiful and full of potential; for one thing, urine is a rich source of phosphate. And if there's one thing the world is going to need very soon, it's a new source of phosphate.

Phosphate contains the element phosphorus, which is an essential component of the genetic material and cell membranes of all living things. It is also an essential component of fertiliser, so it's fair to say that our food production depends on it. Unfortunately, phosphate mining is very energy-intensive and produces a lot of waste. What's more, only twenty per cent of the phosphate mined actually reaches the plate in the form of food – a staggering eighty per cent is lost along the way, due to inefficiencies in the mining process, in the manufacture of fertiliser, in its application to fields and even in its uptake by crops.

But of even greater concern are forecasts that we may hit 'peak' phosphate in just thirty years.[4] This prediction has already had consequences for the global phosphate markets, which have seen the price of phosphate rock soar from US$50 per tonne to US$350 per tonne in just fourteen months. This rise has coincided with a massive increase in demand for fertiliser for the production of crop-based biofuel.

And yet a solution can be found right under our noses (or our toilet seats). According to Dana Cordell from the Institute for Sustainable Futures at the University of Technology Sydney, urine has a lot going for it as a fertiliser. 'Firstly, it doesn't require any treatment before reuse, as it is essentially sterile,' says Cordell. Secondly, it contains all the right nutrients for food production – phosphorus, nitrogen and potassium – and in the right proportions, too. And thirdly, it's easy to use. 'You can use urine directly – you don't need to rely on extensive and energy-intensive technology to reuse it,' she says.

Some countries have seen the writing on the toilet wall and are investing in technologies to capture and use this waste stream. In some rural parts of Sweden, all new homes are required to be fitted with special NoMix composting toilets, which separate urine from faeces, then divert the urine to a storage tank and the faeces to a composting tank. Householders can then use the urine directly on their gardens as fertiliser or, as happens in some cases, set up arrangements with local farmers, who collect the urine for use on their crops.

As well as extracting valuable resources from waste, NoMix toilets also require much less water. Additionally, with more households using them there's less sewage going to treatment plants. However, Cordell says the biggest challenge will be working out how to capture and reuse urine from the most intensive source: cities. 'The challenge is that while we're producing all these nutrients in the city ... they are needed in food-production areas, which can be quite far away from cities,' Cordell says. 'How do you get those nutrients to agriculture – or do you bring agriculture to the cities?'

One solution is to find a way of extracting the nutrients from urine at the sewage-treatment stage. Commercial operations are already under way in several countries to extract the phosphate in the form of ammonium magnesium phosphate, or struvite, which can be used as fertiliser. Researchers in Switzerland are exploring ways of extracting other vital ingredients from urine, such as nitrogen and potassium, which can also be used on crops.

As the value of waste is widely recognised, more new technologies such as these will emerge. Waste that is not sold will be seen as a lost revenue stream simply given away for free. But knowing what to do with waste is only half the challenge. You have to find it first.

How do you find the waste?

Waste is everywhere. Our houses leak hot air in winter, our DVD players consume energy even when they are turned off, our car engines spew most of their fuel energy out their exhaust pipes ... the litany of waste woes is almost endless.

But how do you find the waste? It's easy – if you know where to look for it. Put this book down for a second and take a moment to look around with your sixth-wave glasses on. How much waste you can identify in your immediate surroundings?

Firstly, let's look at the energy being consumed in the room. Are all the lights that are on totally necessary, or could natural lighting do the job just as well? This is particularly important because incandescent light-bulbs actually produce far more heat than light – in fact, ninety per cent of the energy consumed by a filament lightbulb is given off as heat. No wonder compact fluorescent lights save so much energy.

What about your electrical appliances – are they really switched off or are they in standby mode? For televisions, VCRs and other electric devices, we have already seen that standby power consumes up to ten per cent of the total energy used in an average domestic home[5] – the annual collective standby power drawn from households in the United States is around eight gigawatts.

The room itself is another area to look for waste. Is it properly insulated, or is there hot (or cold) air escaping through walls or gaps in doors or windows? What sort of materials are the walls made out of, and what will happen to them when the room is demolished? How often do you clean the room? What cleaning products do you use, and how much water is involved? Even the roof on your house is a potential source of energy that is rarely capitalised on. It absorbs and radiates heat that could be captured, solar panels could be installed there, or a rooftop garden could thrive in the warmth, with the added advantage that it would keep your house cool in summer. Some businesses are starting to take advantage of the unused space on roofs and are leasing this space from households and businesses to install solar energy generators.

Still it goes on. How much waste was produced in the manufacture of the seat you are sitting on? Where did it come from and how did it get to you? What will you do with this chair when you decide to buy a new one? In fact, there are so many sources of waste in our immediate environment alone that a better question might be: how do you avoid waste?

One way of tracking down and eliminating is to look out for the 'four horsemen' of waste: heat, sound, light and pressure. Many of the most inefficient systems generate more than one of these types of waste. For example, the internal-combustion engine in your car wastes considerable amounts of heat and, when compared to the electric car, a lot of noise (which is a form of energy). Coal-fired power stations waste incredible amounts of heat and pressure.

Waste can also be found in things that are not being used efficiently. Again, examples can be found in your home. If you are reading this book you are probably not mowing the lawn. Your lawn mower is a resource that is not used 100 per cent of the time; this means that the resources that went into building your mower – including the money you pay to service and maintain it – are not being put to maximum use. Some smart businesses have recognised the opportunity in this, and are starting to hire out expensive appliances, from lawn mowers to carpet cleaners.

Then there is the most obvious waste of all: the stuff we throw into our garbage bins, flush down the toilet and dump in landfill. Here the challenge is not so much finding the waste but measuring, mapping and describing it. It's incredibly important to do so, according to UK waste expert Peter Jones. 'It has become evident in the waste industry that if you don't understand the upstream factors that are influencing composition and tonnage flows and the origins of waste, you can't really take a sensible view of the investment strategy when the government decides to move to an improved environmental agenda away from landfill,' says Jones, who is a member of the London Waste and Recycling Board.

Jones is a strong advocate of *material flow mapping* – knowing exactly what waste is going where, when and how. He likens it to how we currently keep track of money. 'We look at money as cash, short-term deposits, gold, mortgages, corporate bonds and so on,' Jones says. 'Money moves through those different types of instruments, and as it leaves one system it enters another but it's all self-balancing.' The same approach needs to be taken for waste. Jones argues that we need a national system that maps the waste flows of major companies, which would account for the movement not only of carbon waste but also of other materials, such as plastic, glass and timber.

Sometimes that movement itself is the issue. Waste can be more than the tangible materials we throw into the rubbish bin. Another major problem that has grabbed consumers' imaginations in recent times is transportation waste.

Stop moving

There's a new beast prowling the aisles of supermarkets and grocery stores these days. It has very specific tastes. Rejecting cheeses from France, grapes from the United States and tinned tomatoes from Italy, it has eyes only for produce grown or made in its local area. Meet the locavore.

Locavores are people who choose to source as much of their food as possible from within their local area. They are the foot soldiers of a growing international movement in favour of local food systems, which has arisen partly out of concern for the quality of fresh produce that has had to be freighted over long distances, but also out of concern for the environmental costs associated with that transportation. 'Food miles' have become almost as important as the calorie count of a product, as communities all around the world launch their own campaigns to 'think global, eat local'. Farmers' markets, where local producers can sell direct to the public, and restaurants such as the GoodLife Modern Organic Pizza restaurant in Adelaide, which sources as many ingredients as possible from the local region, are becoming ever more popular.

Transportation of goods is a major source of waste, and it is particularly bad because the transportation does not add any value to the final product being sold (other than taking it to another place). If anything, transporting fresh food over long distances can reduce its value, as it often means that produce has to be picked unripe and stored for long periods of time. And it's not just food – vast amounts of resources and people are travelling far beyond their original homes, and the environment is the loser. Global carbon dioxide emissions from the transportation sector make up of fourteen per cent of total world emissions. The proportion is even higher in developed countries; emissions from transportation in the United States are about twenty-seven per cent of the total. One flight from New York to Los Angeles and back emits about 2.4 tonnes of carbon dioxide. That one flight produces about twelve per cent of the average US resident's annual carbon emissions.[6]

Transportation inefficiencies aren't found only at the international level; we transport resources inefficiently even in some production processes. In the book *Natural Capitalism,* the story is told of how engineer Jan Schilham was asked to examine the design of a factory in Shanghai. Specifically, he looked at ways to improve the efficiency of the pumps and pipelines in the factory.[7]

Schilham found that the normal design practice for pipelines was to use cheap, thin pipes, and to lay them out around all the other equipment and machinery in the factory. While this was logical, it meant that many pipes were at right-angles, or were longer than they needed to be. A right-angle might look neat but it increases friction within the pipe, which meant that the factory's pumps needed to be larger and used far more energy because of the friction.

Schilham redesigned the system by laying out the pipelines first. He used large-diameter pipes and diagonal layouts whenever possible, which reduced the pressure required in the pipes. This allowed the factory to use a much smaller and cheaper pumping system, and to get rid of eighteen pumps. The end result was that the redesign cut energy consumption by ninety-two per cent. It also created a system that was simpler, cheaper, more efficient and had a longer life.

What Schilham found was that sometimes it is necessary to think differently in order to design the waste out of a system. Something as simple as minimising the transportation cost of resources can turn a system on its head, for the better. But minimising waste is one thing – eliminating it altogether is a far greater challenge.

Closing the loop

Cancer is a horrible disease. Uncontrolled and unwanted cell proliferation upsets the body's natural balance, causing illness, debilitation and sometimes death. We don't always know what causes cancer – sometimes it's an external toxin or infectious agent, other times it's a genetic abnormality. It's often hard to see anything positive about cancer, but Swedish oncologist Dr Karl-Henrik Robèrt noticed how cancer seemed to unite families, friends, the medical profession and the community in a common goal – to pool their resources, compassion and know-how in order to defeat the disease.[8]

It could be said that our planet is experiencing a similar form of unrestricted growth, yet in the late 1980s Dr Robèrt was struck by the lack of such a coordinated purpose when it came to dealing with our environmental problems. The solution he came up with was 'the Four Principles of Sustainability' – principles that must be adhered to if we want to 'maintain the essential natural resources, structures and functions that sustain human society'.[9] Dr Robèrt's principles are based on the observation that in a sustainable society, nature is not subject to a build-up of substances extracted from the Earth's crust by society, or to degradation by physical means.

His principles, therefore, boiled down to four main directives. Firstly, don't take things out of the earth that will build up in the environment. Secondly, don't put things into the natural environment that can't be taken back out. Thirdly, avoid any one-way chemical reactions or processes that cannot be reversed. Fourthly, support the basic rights of humans.

Taken together, the first three principles map out a closed-loop system. This is very much the end goal for waste reduction: the waste output of every process is the feedstock for another process. We've already seen several technology companies that are aiming to develop closed-loop systems: Chris Bosua's EARS device closed the loop in pneumatic tools by recycling waste energy and pressure back into the system; John White's UR-3R recycling system recaptures 100 per cent of the urban waste stream; and Close the Loop's cartridge-recycling process ensures that every bit of a toner cartridge is recycled or reused in some form.

But what about the third principle – no irreversible chemical reactions? This is where the emerging field of *green chemistry* comes in. The idea of green chemistry came in response to concerns about the hazards associated with so many chemical processes; it looks for alternative chemical processes that don't involve so much toxicity and waste. As Professor Roy Jackson of the ARC Special Research Centre for Green

Chemistry at Monash University describes it, 'environmental chemistry is determining the mess and how to clean it up, and green chemistry is making sure there's no mess in the first place ... If we were totally successful, all the environmentalists would lose their jobs.'

Green chemistry is about creating a full cycle of chemical processes, says Jackson, so all the materials involved in making a product are able to be recovered and reused. Twelve principles form the foundation of green chemistry; for example, that synthetic methods should be designed to maximise the incorporation of all materials used in the process in the final product, that chemical products should be designed to preserve efficacy of function while reducing toxicity, that any raw material or feedstock should be renewable rather than depleting, and that chemical products should be designed so that once their purpose is served, they break down safely and don't persist in the environment.[10] Green chemistry research has seen the development of 'supercritical' carbon dioxide as an alternative to organic solvents, and 'supramolecular' chemistry, in which reactions can occur between substances in their solid states, removing the need to dissolve them in solvents first.

One industry where green chemistry has its work cut out for it is in pharmaceuticals, where huge amounts of often toxic materials and chemicals may be used in the production of just a few kilograms of end product. 'Many of them involve lengthy syntheses, with several steps in each one, using solvents and reagents that are not very nice to use,' says Jackson. On average, across this industry the production of one kilogram of end product leads to the creation of anywhere from twenty-five to one hundred kilograms of waste. Green chemistry has created alternatives to many of the most wasteful steps, such as using water instead of solvents or replacing solvents with less toxic alternatives.

Even the famous 'little blue pill' – Viagra – has gone green, winning several environmental accolades for manufacturer Pfizer. The company changed several aspects of Viagra's manufacture, getting rid of tin chloride and hydrogen peroxide from the process, for example, and carefully tailoring the amount of a particular solvent used and substituting another ingredient to reduce carbon-monoxide emissions.[11] The result was that Pfizer managed to reduce the amount of waste created per kilogram of Viagra down to just six kilograms.

Green chemistry, like so many other environmentally responsible endeavours, not only reduces waste but also reduces cost, so it goes without saying that industries applying these principles will not only start to close the loop on their resource usage, but will also operate more efficiently. So why isn't industry falling over itself to implement these efficiency-improving, planet-saving measures?

What are we waiting for?

Energy efficiency might be impressive on paper but it's not very sexy stuff. Ask investors interested in clean technology where they want to put their money, and it's most likely to be in high-tech renewable-energy solutions that gleam and sparkle, like solar cells or wind turbines. Redesigned plumbing or using energy-efficient light bulbs don't quite have the same pizzazz. For this and many other reasons, market forces on their own haven't worked very well in encouraging energy efficiency.

It's not just lacklustre markets that are impeding progress towards a more energy-efficient society. A recent report by McKinsey & Company about energy efficiency in the United States identified a number of barriers to progress.[12] The first is that improvements in energy efficiency usually require a lot of upfront investment, while the payback is incremental and happens over the lifetime of the device or the measures taken. Usually, there are no quick and easy returns from investing in energy efficiency. The second barrier is that energy-efficiency gains are extremely fragmented, being spread out over a huge number of devices, locations and situations; thus, there is rarely a simple and elegant solution. Unfortunately, this means that there can be no single champion of energy efficiency; achieving significant gains requires buy-in from everyone, from the individual householder to the engineer to the CEO.

Because gains at the individual level are so small, improving energy efficiency isn't necessarily going to be high on many people's to-do list. If you're strapped for cash, you won't fork out the extra few dollars for that more efficient lightbulb, even if it will save you money over time. Likewise, some companies can't afford to install more efficient equipment because they have already committed their investments to the existing infrastructure. Often improvements in energy efficiency require investment from people who won't necessarily benefit financially; for example, there is little incentive for the owners of apartment buildings to install more efficient air conditioning if it's their tenants who pay the energy bills.

There's also a vast lack of knowledge, even in business, about how and where energy is consumed. The adage 'if you can't measure it, you can't manage it' is particularly relevant here – most companies are blissfully unaware of the small but significant amounts of waste their buildings and operations leak, because it is very difficult to track and quantify those leaks.

Another difficulty is that plugging leaks often requires behavioural change, and humans are, unfortunately, creatures of (bad) habit. We know that leaving the lights on when we're not in the room, or leaving the tap running while we clean our teeth, is wasteful and costs both us and the environment, but we're not very good at acting on this knowledge. It's even harder when the people doing the wasting aren't paying the bills. Try encouraging people in an office to remember to switch off their computers at night instead of letting them go into sleep mode.

Even consumers who are environmentally aware and sympathetic do not necessarily put their money where their consciences are. An Imperial College Business School study found many otherwise environmentally aware individuals do not subscribe to green power because of factors including lack of personal relevance, inconvenience of switching, uncertainty about green power and a lack of accurate information.[13]

So why not find an alternative that makes use of the waste without requiring any behavioural change? The opportunities in waste aren't always in plugging leaks. Sometimes, the leaks themselves can prove useful.

Share the load

When it comes to looking for the proverbial needle in a haystack, you can't go past the Search for Extra-Terrestrial Intelligence (SETI). Thousands of galaxies, millions of planets and a massive amount of radio bandwidth are being scanned for any hints of intelligent life, but progress is slow because of the sheer volume of data. It would take a supercomputer with powers beyond our current technological ability to scan even a portion of the available frequencies and analyse the results.

No such computer exists; even if it did, SETI's funding would hardly stretch to buying it. However, SETI has tapped into something else instead – the spare processing power of all the computers across the planet. Launched in 1999, S ETI@home is the largest distributed-computing project in the world. Nearly six million users around the world have downloaded special software onto their internet-linked computers that, when their computer is in screensaver mode, processes a tiny amount of SETI data to look for patterns that might suggest an intelligent extraterrestrial signal.[14] The SETI@home network is now the largest and most powerful distributed computer on the planet, capable of performing nearly 700,000,000,000,000 calculations per second; it's almost as powerful as the third-most-powerful supercomputer, IBM's JUGENE.

Since SETI@home blazed the distributed-computing trail, similar approaches have been developed to aid scientists in developing climate models and model protein folding, to great effect. Distributed computing uses wasted computer processing power – the power that's sitting there waiting to be used while you get your cup of tea, attend a meeting or procrastinate by doing the laundry instead of working.

A waste-free world

What does a world without waste look like? It might be difficult to picture but, thankfully, here's one we prepared earlier – nature. Mother Nature has been operating as a closed loop system since the birth of our planet. In natural systems, nothing is wasted and everything is a feedstock for something else. A plant takes nutrients from the ground and grows fruit. The fruit is a feedstock for an animal, and when that animal dies its body becomes food for invertebrates, such as the humble earthworm, which return nutrients to the soil to enable the growth of plants.

Humans are part of this natural system. As organic lifeforms, we are actually very efficient, and as long as we don't allow our waste to build up we can live quite naturally. But unlike many other natural systems, we also create tools to help us do things.

In a world without waste we end up with two types of products. Things we consume and things we use. It is inevitable that the things we consume – like food, fuel or fibre – will become part of a closed-loop system, because if they don't, we'll eventually run out of the resources necessary to produce them. But what about the things we use, but which don't necessarily involve us consuming something – like our mobile phones, our office furniture and our household appliances? They have a different future – one in which the service, not the product, is the name of the game.

Chapter 7

Sell the service, not the product

There's an old joke that says a leisure boat can be described as a hole in the water into which you pour money. Something similar could well be said about a car. To begin with, you spend anything from a couple of hundred dollars to a couple of hundred thousand dollars for your steed. Then there are the fuel costs – thousands of dollars per year, for a typical car doing average mileage. Then there's the annual registration, insurance and servicing costs, and the occasional cash splash to get new seat covers or a sunscreen, or to run it through a carwash. All up, you likely spend several thousands of dollars on your car each year to keep it on the road and running smoothly.

But it's likely that you have only used your car for a short time today, if you used it at all. This incredibly costly resource spends most of its life sitting idle, all the while drawing a constant stream of money from your wallet. A more sensible approach would be to have a car that is there when you need it; when you're not using it, someone else can. Imagine a system in which you don't pay for the car when you're not using it. What you need are all the services that a car provides, without actually having to buy a car.

This is exactly what car-sharing services like GoGet provide. Say you do almost all your travelling by public transport, but every once in a while you need a car to go somewhere out of the way, or to transport something that can't be carried on the bus or the train. Instead of buying a car for those infrequent trips, you become a member of GoGet and are issued with a special swipe card. When you need a car, you can book one online as little as five minutes in advance, and the website will direct you to a car parked nearby. You can even specify whether you need a small or large car. You find the car, swipe your card to unlock it, disable the security system and log your usage, find the key inside, start it up and drive away. At the end of each month you receive your activity statement and bill, much like for your mobile phone.

You don't need to refill the tank with petrol unless it drops below one-quarter, in which case you are asked to refill it at the company's expense. You don't pay for insurance, registration, servicing, cleaning or parking (except when you're using the car), and most importantly, you don't have to buy a car. You pay for the service the car provides – transportation and mobility – rather than for the car itself.

GoGet co-founder Bruce Jeffreys says the idea for the company was born out of personal need. 'We were looking for a solution for ourselves as well,' says Jeffreys. 'We were living in the inner-city ... We love where we live but the problem is it's overrun with cars.' And the problem with cars is that they take up space, they pollute, they're expensive, they're noisy, and there are far too many of them.

'The car ownership model is so ridiculously inefficient on every level,' says Jeffreys; this is especially true when it comes to the cost of owning a car. 'It has a very profound impact on people's lives. The cost of running a car roughly equates to a day a week for the average income – it is the difference between somebody working five days to somebody working four days a week. It's not just about cost – it's about priorities and time.'

There is a strong indication that popular opinion about car ownership is changing, particularly among the younger generation, for whom car ownership is no longer such a status symbol as it once was. 'Young people are deriving status from mobile phones and online networks, and cars are not a big driver of status,' says Jeffreys. This is also reflected in a massive decline in the rate of young people applying for their drivers' licences. The recent spate of bankruptcies and bailouts of some of the world's biggest car manufacturers has taken some of the shine off the car industry, he says.

So Jeffreys and fellow co-founder, Nic Lowe, saw an opportunity. 'It was about offering a service,' Jeffreys says; that service benefits not only individuals, who avoid shelling out to own a car, but the environment as well. 'One car of ours is about seven private cars, so that's seven private cars that haven't been purchased, so we're slowing consumption down.'

This shift, from product to service, might not seem like much at first glance. But it will change the face of commerce in the sixth wave.

The biosphere versus the technosphere

Car-sharing is an example of the second big idea of the sixth wave: that, ultimately, everything that we use will become a service. Selling the service of mobility rather than the whole vehicle is good for everyone – the consumer of the service gets to share the running costs with others, the company spreads its fixed costs over a number of consumers, and the environment loses fewer natural resources to provide the same utility.

The distinction between the things we *consume* and the things we *use* is a fundamental one for the sixth wave.

We use a car but don't consume it. We consume food rather than simply using it. The world of consumables and the world of technology and tools have become known as the *biosphere* and the *technosphere* respectively, and both have a very different future ahead of them.

The biosphere is the home of consumables – people, animals, trees, insects and all things tangible. It is a world of consumption, where we drink water, eat proteins and starch, and consume heat derived (largely) from the sun's energy. The technosphere is a very different world. It is filled with products that make a difference in our lives but that are not necessarily consumed in the process, such as toothbrushes, air conditioners and microwave ovens. These are the things that make us comfortable by performing a particular service, or that help us interact with one another through travel or communications.

Dividing our world into the biosphere and technosphere gives us greater insight into why we are consuming resources unsustainably. In the biosphere humanity has a reasonably small footprint (the total biomass of humanity is one-tenth of that of ants, for instance) and our waste products are, in theory, entirely recyclable. However, we have allowed a whole range of things in the technosphere to drift into the biosphere, things that consume resources and produce waste products.

So what does a resource-limited world look like? To consume the smallest amount of resources possible and produce the minimum amount of waste, everything in the biosphere – the things we consume – will trend towards being fully recyclable, or closed-loop. On the other hand, everything in the technosphere – the things we employ but which do not necessarily involve irreversible consumption of a tangible resource – will trend towards becoming a service.

The move towards providing a service rather than selling a product has already happened in many areas. Think of the aeroplane, for example. Unless you are loaded with cash, if you want to fly from one continent to another you don't go out and buy the latest Gulfstream aircraft. Instead, you pay for the service that will take you from one place to another. It's a far more efficient way of using a very costly resource; by becoming a service, aeroplanes can be used by far more people, far more often, than if they were only ever owned by private users. And because the companies operating that service make money from the service, not the product, there is a strong incentive for them to be as efficient as possible in providing that service – by fitting in as many passengers as possible, for example, or by restricting luggage weights and only flying profitable routes.

The mobile phone is another product that has become far more profitable as a service. When mobile phones first came out they were very expensive – they still are, if you happen to lose one and need to replace it in the middle of a contract. But when mobile phone companies realised that the real value in the phone was not in the hardware but in the service – the phone call –

they started giving away the handset very cheaply or for free.

As we will see, one of the most profound institutional shifts of the sixth wave will be when businesses and economies recognise that everything we *use* in the future will tend to become a service, not a product. As resources become scarcer, businesses and economies will shift away from being product-based to being service-based. For many companies, this shift will require as fundamental a change to their business models as the internet required in the fifth wave. In this new world, the most successful will be those who work out how to sell the service, not the product.

Service sells

As we saw earlier, IBM is a rare beast – a company that has managed to survive and thrive across several waves of innovation. Having ridden the fifth wave in spectacular fashion with its manufacturing of personal computers, it is now reinventing itself to surf the sixth wave. This time, though, IBM is shifting its focus from products to services.

In the 1980s and 1990s the company sensed a change in the wind, says Glenn Wightwick, chief technologist for IBM Australia. 'We absolutely recognised ... that there was a requirement to build on what I'd call our systems business, which is hardware and servers and so on, to build a software practice or portfolio on top of that.'

That included products for the management of data and transactions, for software engineering, for system management and for collaboration. Then there was also the service of making all the software work – operating it, building it, testing it, delivering it and managing it. 'That was critically important, and that drove the whole explosion in the services business, to the point today where services, for IBM, is about fifty per cent of our business portfolio,' says Wightwick.

As a global company, and thanks to the developments such as the internet, IBM is well positioned to deliver these services. 'The internet has played an enormous part in allowing us to deliver services with resources located around the world,' Wightwick says. 'We've built expertise in all sorts of different domains in all sorts of different parts of the world ... the ability to collaborate and connect and engage all of that together to solve a problem is really what it's been about.'

What IBM realised was that successful firms were increasingly shifting focus from products to services. So shaking off over 100 years as a hardware manufacturer, the firm made a radical step towards the world of services, selling its hard-disk operations to Hitachi in 2002 and its PC manufacturing industry to Lenovo, and at the same time acquiring the consulting arm of professional-services firm PricewaterhouseCoopers.

Many other companies are now following IBM's lead. Apple made a huge splash when it launched the iPod, but on one level the iPod is simply a hook – a ridiculously profitable one, admittedly – that pulls you into Apple's iTunes service. While the amount of money Apple actually makes from iTunes is a closely guarded secret, there's little doubt that its financial star is on the ascendant as more people switch from buying CDs to buying music online, and as an increasing number of devices tie into iTunes as an online retailer.

And this is only the beginning. Focusing on services creates many opportunities that are not available in a product-based world and, increasingly, the boundaries between products and services are blurring. Customisation and personalisation become hooks into a service, as firms realise that services add value to a product. For many businesses, services are a new way of thinking; when creating something new, they must think not about the product they're building but rather the service they will deliver.

Put simply, services are interactions that create and capture value.[1] However, they are often given a more simple definition: services are anything that you can't drop on your foot. They deliver help, utility or care, an experience, information or other intellectual content from one entity to another. They can be completely independent of the physical world, and their success depends much more on having better intellectual property than the next guy, rather than on consuming more or better resources.

While services might be at the heart of the sixth wave, the shift away from a product-based economy and towards a service-based economy has already been happening slowly but relentlessly in the economies of many developed countries for the last fifty years. At least seventy-five per cent of the GDP of major industrialised nations is in the service sector. The sector itself is worth US$2.4 trillion in exports, and this figure has increased by fifty per cent since 1970.

If you scratch below the surface of many resource-based economies, you'll find even more services. For example, twenty per cent of the added value in Australia's manufacturing, rural and mining exports is in selling services created during the mining process – prospecting services, for example. It's no longer enough just to sell a product in the same way you've been doing it for the past decades. Increasingly, traditional goods-based industries need to offer some extra service to compete in global markets. If this concept of a service-based economy sounds alien, it's actually all around us and we're totally dependent on it. Financial services look after our money, public transport services get us from A to B, utility services ensure we have power and water – services are critical for the smooth functioning of all businesses and governments.

But what is most important in services is that most of the value is intangible, rather than residing in any physical product. This makes services particularly well suited to the sixth wave of innovation. Instead of everyone buying a car, with all the associated resource consumption and waste, in the next wave we'll have an alternative scenario in which there are fewer cars and everyone subscribes to the service of transportation. Fewer resources will be consumed to meet the demand, and those resources will be managed centrally, and more efficiently. Success in the sixth wave will be defined by how effectively businesses and nations move from production to services as a source of competitive advantage.

A services kind of world

For many businesses, services represent a new way of thinking. Instead of deciding what type of product they are going to build, they have to work out what type of service they will deliver. Car-sharing, air travel and software are fairly obvious examples of service-based industries, but carpets? Short of inventing a flying version, how exactly can a carpet become a service?

The global carpet manufacturer Interface is a much-praised pioneer in sustainability, thanks to the foresight of founder Ray Anderson. He realised that the carpet Interface made required a great deal of non-renewable inputs from the biosphere, and that far too often these ended up in landfills or incinerators. Ray started by examining the resources that went into making carpet. By identifying and eliminating many of the waste streams associated with carpet manufacture, and by developing a comprehensive recycling and reuse program, Interface succeeding in reducing the amount of water and raw materials it used by up to seventy-five per cent.

This was ground-breaking stuff, but then Interface took an even greater leap into the unknown. Anderson looked at the most basic premise of the company's business model – that it sells carpets – and turned that on its head. Rather than selling carpet tiles to customers, Interface decided to rent them.

The Evergreen Lease program allows customers to choose whatever style carpet they like, but instead of buying carpet tiles outright, they pay a monthly rent. As part of the lease arrangement, Interface replaces any carpet tiles that get damaged or worn, which means they are then able to 'repurpose' them into new carpet.

Effectively, what Interface did was to break away from selling carpet; instead, it decided to sell something quite different – the *service* of the look and feel of carpet. Clients like it because it's easier on their cashflow.

Interface profits because focusing on the service has allowed it to ensure that prices of its products accurately reflect their true value to the consumer, and reusing materials has enabled it to increase its market share at the expense of its more inefficient competitors. The environment likes it because much less carpet ends up in landfill.

Interface's new business model – part of what founder Ray Anderson calls 'redesigning commerce' – also demonstrates something even more fundamental: services align incentives between the producer and consumer. Both want the service for as little resource as possible. By focusing on the service being provided, Interface has brought its needs into alignment with the needs of its customers.

Aligning incentives

You could argue that there is a lot of waste produced today because of a mismatch between the goals of producers and consumers. Consumers want the products they buy to last as long as possible. Producers have a different incentive – they want their products to last as long as they can reasonably be expected to last, but not longer. Otherwise, the market for these products in the future is reduced. Known as *planned obsolescence,* this is part of many products' design criteria and is responsible for a great deal of waste. It means that many products are being designed – deliberately or inadvertently – to end up as landfill sooner rather than later, to make way for their successors.

In the services world, the misaligned incentives that drive planned obsolescence disappear. Both the producer and the consumer want the service to last as long as possible, as it cuts the cost down. Manufacturers have an even greater incentive to make a durable, long-lasting product because they, not their customers, bear the cost of replacing that product when it wears out.

As an example, imagine if tyre companies only rented tyres rather than sold them – effectively selling tyre services. This would create an incentive for manufacturers to develop a tyre that would last for 200,000 kilometres rather than 50,000 kilometres. Tyre companies might also become more involved in their product, contracting service stations to check tyres regularly and ensuring that they are healthy and robust. They may even start to monitor your tyres remotely to ensure that they are at optimal pressure, as this will save the company money by reducing wear on the tyre.

The services approach has another highly desirable effect on how manufacturers view their products. If they are maintaining ownership of a product throughout its life rather than handing over all responsibility for it to the product's new owner, there is a strong motivation to think about what will happen to it at the end of its useful life. It's hardly going to look good on a company's balance sheet if it is burdened with tonnes of unusable, toxic or otherwise useless waste at the end of it all. This creates a very clear incentive for the company to consider the entire life cycle of its products – for example, how to break them down and reclaim as much as possible of the resources that went into their creation.

Take photocopiers, for example. Companies that rent their copiers to clients and charge them out at a rate per page are more motivated to think about what happens when they get the photocopier back at the end of its life. They will make it modular, so that it's easier to break down and separate its components. They will want to ensure that as much as possible is recycled, and they may even find ways of reusing certain parts directly in new photocopiers.

Taking a services approach to commerce not only makes sense in a resource-limited world, but it also highlights the fact that, in the technosphere, the value of things is in the service they provide. What is important is to separate the value created in the technosphere from the cost of the resources consumed in the biosphere.

This trend towards services isn't just occurring in businesses that previously sold products. Many parts of the resources economy are also realising that what they really provide is a service, not a commodity. For example, might energy be viewed not as something that we consume, but as a service?

Consumables as services

What is fuel? Taken at face value, fuels such as oil, petrol and kerosene are essentially a collection of carbon and hydrogen atoms arranged in such a way that when you introduce oxygen into the mix, you get a reaction that generates heat, water vapour and carbon dioxide. These hydrocarbons are highly sought after all around the world, and humankind has spent a great deal of time and money looking for them.

But let's look at fuel from a services perspective. Do we really want to own litres of oil, tonnes of coal or megawatts of electricity? The answer is no; what we are looking for is the services that the fuel enables – heat, light and mobility – and these are services that can be sold just like any other, and which preferably don't consume any natural resources.

The UK company Thameswey is one of a new breed of energy-services companies (ESCos) that have designed their business models around selling energy services rather than energy. What sets Thameswey apart from most other energy companies is that it doesn't just sell electricity, but it sells the service of heat as well. 'Normally when you generate electricity, a lot of heat is generated and that's wasted,' says John Thorp, group managing director for the Thameswey group of companies. However, Thameswey has found a way to turn that wasted heat into a service. Its electricity-generation mechanism is a modified marine diesel engine that runs on gas instead of diesel, and it has set it up as a combined heat and power system. 'In a combined heat and power system you capture the heat that is normally wasted and you distribute that around a district heating network,' Thorp says. That makes the conversion of fossil fuel to usable energy very efficient, giving Thameswey an efficiency rating of above eighty-five per cent.

It also means it is able not only to sell the electricity it generates but also the heat. 'We collect that heat in a thermal store and then we distribute it through insulated pipes to the buildings that we supply,' Thorp explains. The heat is distributed via insulated pipes and heat exchangers. Customers control the temperature and the heat supply is metered. 'It works out cheaper for the building owners because they don't have all the investment in maintenance of equipment that you would normally have if you had boilers in your building.'

The Thameswey group is unusual in that it is a wholly owned subsidiary of the Woking Borough Council. It was conceived in 1992 as part of the council's climate-change strategy, which aimed to improve the efficiency with which energy, heating and cooling were supplied to corporate buildings in the town centre, and, at the same time, to reduce the council's CO_2 output. The system now produces 1.3 megawatts of electricity and heat in Woking, which supplies more than twenty buildings, and six megawatts in nearby Milton Keynes, supplying four major buildings.

Thorp sees combined heat and power systems as the way of the future. 'It has to be a model for the future because fossil fuel prices are going to rise enormously as they become constrained, and therefore the more efficiently you can convert the fossil fuel into usable energy the better.' Combined heat and power units can also be built close to the end-user, cutting down transmission losses and distribution costs.

In France, companies called *chauffages* have been doing this for quite a while. A *chauffage* contract works by taking out a contract of twenty to thirty years with a service-provider to heat a building.[2] Such an agreement creates a clear incentive for the *chauffage* to provide its service while consuming as few resources as possible, and the contract might even include free insulation for the building.

In fact, the idea of selling energy services rather than energy itself actually goes all the way back to Mr Electricity himself, Thomas Edison, who initially sold light by the 'lighthour' rather than by kilowatt hour.[3] In his initial model, the company that delivered the energy sold the appliance as well; as the number of electric appliances grew, the company decided that its energy and appliances divisions should be separated permanently. Edison General Electric became a commodity seller, and the appliance division became General Electric. Until this happened we were very close to a service model for the sale of household appliances.

Selling services in this way changes a company's business model so that it charges according to the outputs – such as heat or light – rather than the inputs, namely fuel or electricity. For example, GoGet doesn't charge customers who use its car-sharing service for the fuel they consume but rather for the kilometres they travel and the time they spend using the car. For years now, oil and gas companies have realised that they are energy companies. But in a new world they may also sell services; in the words of one executive, 'People will always want light, heat and mobility, they just won't want the carbon downside.'

How would this all work in practice? Your energy bill would probably look very different from how it looks today. Instead of charging per kilowatt hour, it might charge you for the number of 'light-hours' you use to keep your house lit – with the lightbulbs thrown in for free. It may charge for keeping your house within a certain temperature range, and also provide the insulation. You may even sign a contract for slices of toast in your toaster – a 'price per slice', perhaps – or the service of keeping food cool in your fridge, where the toaster or fridge are rented to you as part of the package.

By measuring and paying for the service provided, you decouple the value of the service from the value of the fuel or the product that produces it. Services make explicit exactly what you are paying for. This is significant, as some energy-efficient devices cost a little more than their less efficient counterparts. The payback is in the future, as the financial benefits from a more efficient product accumulate over time. Some people might not be in a position to pay for this initial investment, or they might be sceptical about the gains to be made; by renting the equipment and selling a service, this disincentive is removed.

This has been recognised in the solar-power industry, where the large setup costs can be a big turn-off for consumers, even though the whole-of-life costs can be smaller. While governments are slowly introducing incentives, such as rebates, to encourage homeowners to install solar-power devices in their homes, a typical photovoltaic array or solar hot-water heater still represents a pretty major spend. So we have the situation of lots of nice, sunny roof areas that could be generating solar power, but aren't because the cost is too high.

US energy company Citizenrē has come up with an elegant solution called REnU. The company will cover the costs of installing and maintaining a solar system on your roof, for which you pay a flat monthly rent. In return, you can use the solar energy generated, and if you use less energy than your solar array generates, the excess is delivered to the grid and you earn energy credits, which you can save up and use on the days the system doesn't generate enough power to meet your needs.

By selling a service, REnU takes away the disincentives that deter many would-be solar users, such as the high costs and complexity of solar system equipment, installation and maintenance. And it's making use of a valuable resource – sun-bathed roof space – which would otherwise be wasted.

A shift to services creates a new relationship between the customer and the provider. Customers for energy and water services become increasingly engaged in the provision of services in traditionally supply-dominated industries, and companies that sell services become more inclined to involve users to reduce their resource consumption. No longer is it in the best interests of the producer for you to consume as much of their product as you can – they would rather talk to you about 'switching it off' and 'turning it off'. Encouragingly for these new utility service-providers, experience has shown that consumers are far more likely to adopt new technologies, such as energy efficient lightbulbs or washers on taps, if the utility provider is involved in the process.[4]

Selling the service and not the product makes a significant difference to many businesses, but it is still challenging. Bundling services together requires not only smart business models but also a total rethink of the way that we pull information together.

Service-oriented software

When it comes to the paragons of inefficiency, there are obvious examples – combustion engines, coal-fired power stations, incandescent lightbulbs and so on. But then there are the less obvious, such as software. The inefficiency in software isn't so much in the material resources involved in its creation or distribution, but in how it is used.

Most people have some sort of desktop productivity software package, such as Microsoft Office, iWorks or OpenOffice. Think for a moment about the dazzling array of services each of the programs in that package offers. Picture the toolbar across the top of a program window – there are options for formatting text, creating tables, inserting footnotes, linking to websites, creating columns, numbering pages ... the list of functions goes on and on. Many of those functions are repeated in the different programs in your package – you can create bold text in your word-processing software, in your spreadsheet application, in your desktop publishing software and in your presentation software.

But how many of these functions do you actually use on a regular basis, or even on an irregular basis? If you're a typical home user, probably not all that many. So, not only are you paying for a host of functions, or services, that you never use, but the code for many of them has had to be written separately for each program in the software package. The scenario holds even truer at the business level, where software packages get bigger and more complex.

This is where service-oriented architecture, or SOA, comes into the picture, as Tim Sheedy, senior analyst with Forrester Research, explains: 'Let's say you have a customer database, and then you've got a CRM (customer-relationship management) system that has to get access to the customer details, some sort of finance system that gets access to customer details, and some sort of customer-service system. The way we write software at the moment is that the concept that will get customer details is written three different times in three different applications. That takes server capacity, it takes processing resources, and we're paying for those resources ... Effectively, you're paying for it three times over.'

Software design is often based around things called *objects.* Object-oriented architectures are concerned with representing these objects in code, their properties and how they interact. But are objects really the best way of thinking about the world? When you drink water from a glass, are you really concerned with how many atoms makes up that glass? Are you worried about the viscosity of the silicon dioxide or the reflectivity of the glass? Not really. What you are concerned about is the service the glass provides (holding your water), how it feels in your hand and looks to the eye.

SOA turns traditional programming on its head. Rather than thinking about the object that is being represented in code, it looks at software from the perspective of the specific services that a user requires. For example, in the scenario above, all three systems have services in common – they all have to extract customer information from a database, they all have to display that information on your screen, and chances are they will all have the capability to add new customer information to the database.

Rather than having those services written three times into the three systems, they are written once and shared between the systems. 'The idea of moving to service-oriented architecture is, first of all, you have a more efficient infrastructure where you get reuse of services,' says Sheedy. The second advantage of SOA is that it strips software applications back to the services you really need and gets rid of the stuff you don't use. 'Often you only really use maybe forty per cent of the functionality, because they provide huge capability out of the box, most of which you never use.'

This approach dramatically reduces the processing capacity used, because instead of running an entire software package you only need to run elements of it. For example, for end–of month financial reports, you might only need three or four of the services in your software package, but you use them very intensively. 'So therefore we give those three or four services a lot of the processing capacity at the end of the month, as opposed to the giving the whole financial application a lot of processing capacity.'

SOA is not yet widespread, but some of the bigger software companies are now releasing software that is SOA-enabled. However, the SOA approach demands a shift in billing practices, as rather than buy an entire software package, now consumers can access a particular software service only when they need it.

SOA works in tandem with a new model of software use called 'Software as a Service', or SaaS, as IBM's Glenn Wightwick describes: 'In the past, and we would still do it, we might sell a customer a licence for software and they might pay us a one-time charge or a licence fee every month for however long they use it.' Under a SaaS model, however, the software company hosts and runs the software package and the customer only pays on a per-transaction or per-use basis. 'This is becoming much more popular because your payment stream as a user of that software matches your use of it rather than you paying for it and not using it as much as you expected to,' says Wightwick.

But the real power of this new way of looking at computing becomes apparent when these services move off the desktop and into the real world. As we have seen, except for things that we consume directly, like food and water, it is possible to think of almost everything in a services way. What happens when all of these services – such as energy, mobility, comfort, finance, commerce and bread-toasting – all start to interact? The move towards services-oriented architecture is enabling a brave new world in which these services are able not only to fit together but also to connect with each other in some exciting ways.

The services internet

Imagine you're planning a holiday overseas. You probably have a number of services that you will want to use to book flights, arrange hotels and travel insurance, reserve hire cars or book taxis, and organise visas. In the past, a travel agency might bundle these services together for you and organise a package deal, where it takes care of and coordinates everything.

In the future this packaging will become the standard way of delivering services – but you'll be able to do it all yourself. For example, when you book your flights the service-provider will work out what visas you need and organise your applications directly with the various consulates. If you book a flight and arrange a hire-car together, the airline will inform the car company if your plane is running late. If you want to travel to an exotic location, your guidebook might be able to reserve you a table at a local restaurant. Your internet service might put all of this together, and also recommend the most appropriate level of insurance cover for your destinations and buy it for you. The key is that all of these services are talking to one another.

The European Union calls this the shift from the *internet of things* to the *internet of services.* [5] The implications are significant. Rather than building applications that pull together a range of data sources (or objects) on the net, companies will build them to pull together services. Users will then be able to find these services, combine them and adapt them to their own needs.

One way of doing this is through web-service mashups. A *mashup* is when someone 'mashes' together a number of different things to create something entirely new. It might be joining an airline's booking service with a visa-application service and a trip planner. Alternatively a company could combine a mapping application with a restaurant guide and centralised restaurant-booking service, so you can use your iPhone or similar device to see what restaurants are in your vicinity, then read reviews of those restaurants and book one, all from the same website.

The key to a mashup is that the company that created the original service makes it available to other people to reuse. Google is particularly generous with its various services. For example, free access to the Google Maps interface is enabling a host of web mashups, like CoffeeSeeker.com , which shows you all the major coffee outlets in your area, or Woozor.com, which mashes together an online weather service with Google Maps to create a weather map of the world.

As it becomes easier to join services together, companies that have shifted to service models will find that it is easy to broaden their markets. Once you sell one service, the door opens to bundle other related services with it. For example, imagine that instead of a car-sharing model, you sold cars like mobile phones. In other words, instead of buying a car, you took out a contract for 'mobility' from a combined fuel and automotive company. You might get the car for free as long as you agreed to pay a certain amount per kilometre, which would cover fuel and basic running costs, tracked by GPS. The large up-front cost for the consumer is removed; in fact, this is the model being pursued by many electric vehicle companies.

A company that sells mobility can then bundle other services together and so create even more value. If you are tracking a car using GPS, perhaps you can bundle insurance with the mobility such that the premium is based only on when and where you drive the car. If it is parked for five days a week in a secure location, why should you pay for seven days' worth of insurance?

Bundling and mashups aren't limited only to the commercial world. Governments are also getting in on the act. In Norway, the government compiles individuals' tax returns, using a whole range of services provided by employers and banks. Citizens are sent pre-completed tax returns for their verification.[6] It's a short hop from here to a situation where you'll be offered boat rental when you apply for your government fishing licence.

A services future

In a world of limited resources, there is a strong incentive to create a service for everything we don't consume. Even some of the things that we do currently consume, like energy, can be thought of as services. This is not just a change in business models, but a whole new approach to commerce.

Services allow us to decouple economic growth from resource consumption. Not only does this improve the natural environment, but it allows service-providers to capture the value that is created when the inefficiencies of large waste streams are reduced.

Services make clear exactly what is being provided. By measuring the actual value to the customer, the inputs to the system become less important. The incentives of the producer and consumer are aligned, which creates an incentive for both to reduce the amount of waste in the system. Finally, the entire value chain of services (and products) in the economy is drastically rethought, as services are mashed up in whole new ways; the internet of services emerges.

If you're in business, take a moment to think about what it is that you actually sell. Now, think about the service that you provide to your customers. Is it the same thing? Can you think more broadly than this? Do you sell insurance or peace of mind, for example? Do you sell lawnmowers or short grass? If you can think about the ultimate service, you're on track to make the sixth-wave shift of selling the service, not the product.

Of course, an important factor in providing a service is being able to keep track of how it's being used and the resources it's consuming. This leads into the next big shift of the sixth wave, in which the measurement and monitoring of services and natural resources will become all-encompassing – so much so that the digital and natural worlds we live in will start to converge.

Chapter 8

Digital and natural converge

In the beginning there was the *bit* – the most basic unit of information capacity. A contraction of the term 'binary digit', the bit can exist in only one of two states – on or off, true or false, long or short, punched hole or no hole, one or zero.

Then eight bits came together to form a *byte.* Why eight? Because that allowed enough combinations of ones and zeros to represent the Western alphabet and numerical characters. Then came the *kilobyte* (1000 bytes), the *megabyte* (one million bytes) and the *gigabyte* (one billion). What next? We are already dealing with *terabytes* (one million million), but soon *petabytes* (one million billion) and *exabytes* (one billion billion) will start to find their way into our lexicon.

As we have seen, this spectacular explosion in processing capacity, which laid the foundation for the last wave of innovation, followed Moore's law, a prediction that the processing power-to-cost ratio for a microprocessor would double around every eighteen months or so. This amazing phenomenon shows no sign of slowing down over the next decade, and will continue to enable new technologies and new services yet to be dreamed of.

But there are two sides to Moore's law. On one hand, it means you can pack exponentially more transistors onto a single microchip for the same price – getting massively more bang for your buck. But the flipside of this is that you can now get the same amount of processing power for a fraction of the cost of just a few years ago – the same bang for a lot less buck. The sixth wave will be based as much on this second facet of Moore's law as on the first. Instead of processing power being found only in giant computers locked away inside hermetically sealed rooms, microprocessors will be cheap enough to be used just about everywhere, in everything from passports to pets.

There are now more transistors produced in a single year than there are ants on the planet[1] – roughly the same as the number of rice grains produced globally each year. In the past, these transistors would have been embedded into microchips that were destined for a life of isolation in a calculator, a car, an alarm clock or a computer. They would have done their job in that particular device, end of story. But with the sixth wave, something new is happening. These processors are starting to communicate beyond the confines of their device, and are sharing information that has never been shared, or even collected, before. This brings us to the third big idea that will govern how we innovate in a resource-limited world – the convergence of the digital and natural worlds.

Instrumented, intelligent, interconnected

Ask people which part of their body they love the most, and you'll get a lot who pick their muscular arms, slim legs, pert breasts, six-pack abdomen or perfectly-shaped bottom. They probably spend a fair amount of time looking after these assets, ensuring they are healthy, clean and well looked after. It's a fair bet than almost no one will choose their feet – which is strange. Our feet are our foundation, yet we give them very little attention, love or care.

That same attitude often extends to our car's 'feet' – its tyres. We often only check or change our tyres when something goes very wrong, like a flat, or when the tyres are so close to the end of their life that they are a safety hazard. They are so important to the smooth running of the car and our safety, but how many of us check the health of our car tyres on a regular basis?

To the rescue of neglectful types comes Pirelli's Cyber Tyre. This 'intelligent tyre' has a microchip embedded in it that not only records basic information such as tyre pressure and temperature, but also more complex details, such as acceleration along three axes, the force of the contact between tyre and road, and the load across the tyre.

The Cyber Tyre uses these sensors to monitor the road condition in real time, which means it can sense when it is in danger of losing traction and warn the car and driver before the tyre starts to slip. It also transmits this data in real time to the car's onboard electronic systems, such as the electronic stability program and anti-lock braking system, enabling greater stability control. It can provide information about the user's driving habits and make recommendations on how to inflate the tyre for maximum efficiency. Best of all, it requires no extra power – it uses the energy from mechanical vibrations as the tyre travels along the road to power itself.[2]

The ubiquity, low cost and processing ability of microprocessors have made it not only possible but cost-effective for Pirelli to use them to collect data about an interaction that until now has been an unknown: what happens at the interface between a car's tyre and the road. Pirelli is creating a digital version of something that previously existed only in the natural world – the tyre.

The digital world and the natural world have remained largely separate, but increasingly the boundaries between them will blur. Every interaction with the natural world, particularly with the tools or appliances that you use, will be measured and monitored. IBM calls this the Smarter Planet; it has identified three key characteristics about this interface between the digital and natural worlds – it is instrumented, intelligent and interconnected.[3]

Our world and the objects within it are becoming instrumented in a fashion that could scarcely have been imagined just a decade ago. One of the technologies underpinning this is the radio-frequency identification (RFID) tag. These are small microchips, barely the size of a grain of rice, which consist of two parts. One is a miniature electronic circuit that can store and process information, and receive and transmit radio waves. The other is the antennae, which are needed to pick up and send the radio signals. RFID tags cost a fraction of a cent and, while some require a tiny internal battery, many operate by stealing a little power from ambient radio waves, much like a wireless did fifty years ago.

Their size, cheapness and simplicity have enabled RFID tags to be used in a vast array of applications. If you have a relatively new passport, it will almost certainly have an RFID chip in it that contains your personal information and a unique identifier, such as a digital picture of you. Electronic toll collection systems on major highways use an RFID tag in the car-mounted device to record your passage through a toll gate. Libraries are replacing barcodes on books with RFID tags, and golf courses are using them to track down lost golf balls. Even pet owners can implant an RFID tag into their beloved to ensure that if their pooch or puss strays from home, a quick scan will reveal its home address and the contact details of the owner.

RFID is also proving useful in the fast-moving world of consumer goods, sending manufacturers information about the movement – or the lack of it – of their products, thus helping them find ways to improve sales. Gillette has led the way in using RFID to learn about when and how their products move from manufacturer to distributor to seller to consumer. In 2003 it started attaching RFID tags to some shipments of its Mach 3 razorblades, which were destined for two Wal-Mart stores. These particular stores were equipped with 'smart' shelves that could read signals from the tags, so when the stock started to run low the shelves would alert store staff to restock them, and would even automatically order more razors from the distributor. The company used a similar approach in 2005 to track stock movement following a special promotion, discovering that if stores made sure the product was stocked ahead of the promotion, sales were much higher than if the stock was put on shelves after the promotion had started.[4]

As more companies work together with large vendors, the RFID tag will eventually replace the barcode as the product identifier of choice. Imagine simply walking through the supermarket checkout and having your purchases tallied up without having to go through the tedious chore of unpacking your trolley, having each product scanned individually, and repacking your trolley. This will be easier for both the supermarket and the shopper, but these tags, and technology in general, are going one step further – they are becoming smarter.

The second characteristic of the trend towards integration of digital and natural is that devices will become more intelligent. It's not enough for a product or appliance to simply be instrumented – they will also know what they are for, what they are doing and where they are. This could be as simple as knowing their current status to being able to receive instructions and process information.

For example, light switches in the future may each have an IP address – a numerical label given to any device that is part of a computer network – which would enable them to know whether they are switched on or off. Lightbulbs could also have an IP address and be able to accept an instruction to illuminate. Home networks will connect the two together so, in the future, instead of having to hire an electrician to physically connect devices with power sources and switches, you'll be able to rewire your house at the press of a button. Parents of mischievous children beware!

Intelligent devices will be everywhere. Vending machines will be able to keep track of their stock levels and report back to a central computer system when they need restocking. Your car will know your location and will be able to download road service information or a software upgrade. Even your toilet will be able to report the number of flushes each day to your water-management software.

The ultimate intelligent device has to be the mobile phone, which is increasingly becoming the digital doppelganger of its owner. Thanks to technologies such as GPS, Bluetooth and RFID, your phone knows where you are, what you are doing, and what and who is nearby. It knows where you've been and where you're going, and even how many steps you've taken to get there. Your phone will be able to interface with all the other intelligent digital devices out there, just as you interact with the natural world.

With so many intelligent devices around, we might even reach the point where people find ways to rent out some of their excess processing power. The American computer pioneer (and the coiner of the term *artificial intelligence)* John McCarthy actually had this idea in 1960, when he suggested that 'computation may someday be organised as a public utility'. This is now a reality in the form of *utility computing,* which is the provision of computing resources, such as storage and processing power, as a metered commodity just like electricity and water.

But intelligent devices do their finest work when they are able to work together; they may one day come to rely on each other in order to operate. This takes us to the final characteristic of connected devices, which is that everything will be interconnected. Ultimately, all of these billions of instrumented, intelligent devices will be able to talk with one another. They will even request services from one another without the need for human involvement at all.

Think of your fridge at home. At the moment, it's a relatively simple tool for keeping things cold – little more than an electric icebox. But in an instrumented, intelligent and interconnected world, it might play a much more complex role. Rather than being a one-trick pony, an intelligent fridge will take a much more active role and become your food manager.

Firstly, the fridge will not simply cool food but will also store it. It will be aware of everything inside it, and all the use-by dates, thanks to information provided by the intelligent RFID tags on your purchases. You'll be able to see immediately what food items are available, and how close they are to being past their prime, via the fridge's external monitor, without losing cold air through the open door.

Your intelligent fridge will interact with other intelligent devices. When you head to your local supermarket for your weekly grocery shop, your mobile phone will sense you are in the shop and will be able to query the fridge at home about its grocery needs. The fridge will assess its contents, compare that with its usual stock and judge which of the usual essentials are missing or running low. Then it'll send that shopping list direct to your mobile phone.

Your mobile could then gently remind you that you have run out of your daughter's favourite type of cheese, or that your own stash of honey yoghurt is so far past its use-by date that it's getting dangerously close to achieving its own form of intelligence. In a fully RFID-capable store, your mobile might even order those items for you automatically; you will simply collect them at the checkout.

And finally, your intelligent fridge will be part of a bigger network of household appliances. This will enable it to be much smarter about how it uses energy. The fridge will be in constant communication with your energy-service company, becoming part of what is being called the 'smart grid' – the unification of the energy grid and the internet.

Smarter energy

Smart grids, also known as the *energy internet,* represent a paradigm shift in how electricity is delivered to the end-user. In simpler times, power stations generated electricity and sent it down a wire to a house, and that was the end of it. Then there was the arrival of off-peak power, which recognised that there were highs and lows in demand on the grid. Smart grids take that idea to a whole new level.

Houses and even individual appliances will talk to the power supply in real time via smart meters, telling it when they need more or less power. In turn, power utilities will be able to closely monitor demand on the network, directing and withdrawing power from districts, houses and even appliances in order to ensure there are no hiccups. Smart grids are also designed to cope with the variability in the supply of renewable energy, ensuring that when we do finally source most of our power from solar and wind, it won't die when the sun goes down or the wind drops.

The US company GridPoint's SmartGrid Platform is one example of how the future of energy management might look. It consists of a network of 'demand-management technologies', the hub of which is called the energy manager. Located in each house, the energy manager communicates with electrical appliances, especially the energy-hungry devices such as air conditioners, water heaters and pool pumps. It not only monitors their activity but can even control them independently to maximise efficiency and comfort – by switching off or reducing the activity of air conditioners and water heaters during peak electricity load times if the appliances aren't needed, for instance. This not only reduces demand on the grid but also saves users money by reducing their electricity usages at times when it's likely to cost them more. It can also give users detailed information about their electricity use throughout the day via a personalised online portal. This can tell them how their electricity use maps onto daily changes in pricing, enabling them to plan their electricity use or alter their pricing plan to reduce costs.

That same information, and access, is enormously useful to electricity utilities. GridPoint's energy manager provides utilities with an incredible amount of valuable information about how, when, where and by what appliances power is being used. But it can also enable them to control appliances remotely in order to manage demand during peak times – it can vary the temperature of air conditioners, for example, by one or two degrees for a brief period of time. Such a change might go completely unnoticed by the occupants of a house, if they are even home, but could make the difference between the grid coping or crashing during a peak period.

Connected to smart grids, smart refrigerators will be able to make decisions about when to switch on and off according to how much demand is being made of the power supply. If the thought of a fridge switching off and all your dairy goods turning an unpleasant shade of yuck fills you with dread, don't panic. Smart fridges take advantage of the fact that a fridge is a very good thermal store – that is, it takes a long time for a switched-off fridge to warm up.[5] So, if a fridge switches off for fifteen minutes, its internal temperature isn't going to change drastically, but that fifteen minutes of unused power might be used elsewhere in the house where demand is greater, or it could remain in the grid to offset demand during peak periods. The same principle could apply to a range of household appliances – for example, smart dishwashers and washing machines that you can load up but then set to switch on during off-peak periods, such as late at night.

The logical endpoint is that our homes, offices, cars and pretty much everything else will be instrumented, intelligent and interconnected. Every device in the natural world will have a digital equivalent. See that games console in the lounge room? This will one day become the digital heart of your home. That mobile phone on your desk? This will represent your digital personality.

The end result will be massive gains in efficiency, as smart grids and intelligent devices manage our resource usage – whether those resources be food or energy – minute by minute, second by second, to ensure that as little as possible is wasted. Energy-management software will be connected to home appliances, security systems and entertainment systems. It will be accessible remotely and will be able to track residents and provide services for them, such as switching on and off lights or unlocking doors.

All these digital concepts and interactions are rooted in the natural environment via the devices that host them. But what about Mother Nature – planet Earth? Will she also have a digital equivalent?

A digital planet

If a tree falls in a rainforest and no one is around to hear it, does it make a sound? At a philosophical level, Zen masters might debate this for years. NASA and Cisco have something a little more practical in mind – a massive global network of sensors to monitor deforestation. If a tree falls, they'll know about it.

When NASA and Cisco get together to talk about saving the planet from climate change, you know that something interesting will come out of it. That something is Planetary Skin, a massive global-monitoring system of environmental conditions. Planetary Skin was conceived after the 2009 World Economic Forum identified a critical need for a 'globally trusted mechanism for measurement, reporting and verification' of climate change–related parameters, such as carbon dioxide emissions and carbon flows.

What you can't measure, you can't manage and, more importantly to business, you can't monetise. While Planetary Skin's overarching goal is to rescue the planet from climate change, there's also a truckload of investment opportunity to be explored. Cisco and NASA estimate that, between 2010 and 2020, Planetary Skin could unlock over US$350 billion per year of investment in infrastructure to support mitigation of, and adaptation to, climate change.

The vision is that Planetary Skin will be a system that can stitch together information from a huge network of sensors, ranging from satellites orbiting the planet to ground-based remote-sensing systems – described as SensorFabric – and create a global real-time picture of carbon stocks and flows. That information will then be analysed and presented in a usable format, using a mashup of online decision-support tools called DecisionSpaces. The result will enable governments, industry and decision-makers to be more informed about climate-change mitigation and adaptation, and to encourage further innovation by entrepreneurs.

The first prototype of Planetary Skin is Rainforest Skin, which is an attempt to tackle the rampant deforestation of tropical rainforests and the contribution it makes to global greenhouse emissions. The idea is that arresting deforestation is the best method to halt – and perhaps even reverse – the growth of greenhouse gas emissions, and the best way to arrest deforestation is to make trees more valuable alive than dead, as carbon sinks. Rainforest Skin will be a network of real-time, highly distributed sensing of carbon stocks and flows.[6]

This is just the beginning. The founders of the project see the Planetary Skin platform as a 'globally pervasive nervous system' that can provide a myriad of information about the condition of the earth's ecosystem and atmosphere – a kind of digital 'skin' wrapping around the natural world.

Planetary Skin is one of the many projects using technology to capture and represent the natural world in digital form. On 14 August 1959 the first satellite images of Earth were captured by the US satellite *Explorer 6.* The image of a sunlit area of the central Pacific Ocean looks like nothing more than a smear of light on a dark background – it's almost as if they forgot to take the lens cap off.[7] Contrast that with the images currently being taken by the satellite *GeoEye-1,* which has the highest resolution of any commercial imaging satellite. *GeoEye-1* is able to distinguish objects just forty centimetres in size – it would be able to tell if you were reading a broadsheet or a tabloid newspaper (although possibly not which one).

Advances like this in satellite technology have given us an unprecedented bird's eye view not only of our planet but also of our cities, neighbourhoods, streets and even our houses. And thanks to endeavours such as Google Earth and Microsoft's Bing, access to that view is no longer restricted to the military – anyone with internet access can now look up a satellite image of their home.

High-resolution images are just one part of the picture. Even more important is their ability to know what it is that they are looking at. Satellites can tell a lot from the visible and non-visible spectrum of the images that they are taking, from the health and vitality of a crop to the size of mineral resources buried deep within the ground. Some satellites are able to penetrate cloud cover with radar, or measure firefronts with infrared.

We are continually unlocking more applications for satellite data, according to Dr Alex Held, team leader of the environmental remote sensing group in Australian research agency CSIRO's Division of Marine & Atmospheric Research. 'People tend to focus on what they see in terms of Google Earth,' says Held. 'For that sort of use you need very powerful telescopes on the satellites, and you really get the shape of the object but you don't get a whole lot of additional information about what they're made of.' In remote sensing, satellites orbiting the earth take images at wavelengths of light outside the visual spectrum, which reveals a vast amount of information.

'We use thermal sensing to look at the temperature of plants,' Held says. 'When plants don't have enough water or when irrigation is not sufficient, plants tend to warm up above average air temperature, so we detect those changes and use that to interpret that they're under stress, for instance.' Other researchers are using wavelengths that tell them about the concentration of organic matter in water – to study the level of sediments and other materials being transported by rivers and dumped on the Great Barrier Reef, for instance.

One of the main uses of remote sensing is in the area of climate science. Held and his colleagues are using satellites to measure concentrations of carbon dioxide in the atmosphere. 'Carbon dioxide gas in the atmosphere has a very specific absorption in a few wavelengths that we use, the same as the ones we use to measure concentration of ozone,' he says. Thermal imaging is also used to measure the temperature of the ocean's surface, which has a major impact on global climate.

The imaging capabilities of satellites have improved dramatically in recent years, but we also have a huge amount of data collected over the past five decades that is providing a valuable insight into how our natural resources have changed over time. 'We have a very nice record now, for instance, of the changes across Australia in terms of land cover,' says Held. 'These things have to be very carefully calibrated and curated in some respects, but these time series of satellite data are starting to show us quite interesting trends across the continent in response to El Nino, droughts, et cetera.'

In the early days of satellite imagery the data was very difficult to manipulate; computers simply couldn't cope with the large amounts of information involved. However, as technology and storage ability have caught up with the satellites themselves, this information can be processed and integrated with other sources of data to give us a far more comprehensive and useful picture of the world.

Spatial information

Imagine you're interested in buying an inner-city apartment, but you want to know more about the local area before you decide to invest. Are there any good schools nearby? Is it a safe area for your kids to play in the street or for you to park your car on the street? Are there any noisy premises nearby? Where are the nearest restaurants and grocery stores? Can you ride your bicycle around easily? In the past, you might have relied simply on the real-estate agent's knowledge, or done a bit of legwork yourself to answer these questions, but this can be time-consuming and you can't always find what you want to know easily.

Now there's a better way. To begin with, you can take a look at a detailed satellite image of the area and see where your apartment-to-be is, relative to things like parks, office buildings or large roads. That satellite image will also be overlaid with a detailed map that identifies smaller landmarks, such as schools, shops, restaurants and bicycle pathways, and it will also give you access to detailed information about those landmarks.

Want to know about the school? With a few clicks on the map you'll be able to access information about the school's academic curriculum, the extra-curricular activities it offers, how it ranks compared to other schools in the area and even how its sports teams perform. Worried about safety? The map can be overlaid with police reports showing problem areas for burglaries, attacks or noise complaints. Love dining out? You'll be able to quickly and easily access reviews of restaurants in the area, see their menus and order a meal. You can even check out the local weather conditions.

All of these technologies are known as *spatial information,* and they're revolutionising everything from tourism to shopping. Enterprises such as NAVTEQ and Google Street View are currently in a race to take millions of photos of cities all over the world, and with smart technology they plan to create 3D images of buildings. This means you can visit New York, eyeball the architecture, drop into Macy's department store, select something nice from their catalogue or maybe even browse through their virtual reality shelves, and order it to be delivered to your home – all without leaving your house.

A picture may be worth a thousand words, but a map is worth a thousand web queries. Google Earth and Microsoft's Bing Maps are the future of the internet in this respect – both companies are battling for the digitisation of the natural world. Google and Microsoft are investing so much because they realise that one way of managing the ever-increasing amount of information on the web is to give it a location. Rather than having a restaurant review sitting out there in cyberspace with a million other restaurant reviews, it will be attached to the restaurant itself via a digital map. Instead of going to a review site, you'll find the review in its logical home. Even web sites will one day be associated with a location, even if it is simply the address of the registered owner.

Once you have this base data, you can do a lot with it. Applications will draw information from a whole range of sources, all geo-referenced to locations. This location information is already becoming available for a lot of data on the web – there are at least 600,000 Wikipedia entries that include latitude and longitude coordinates, and this number is growing rapidly.

It's not just mapping technology that is improving. We're also developing new ways of collecting information about the world around us and incorporating it into the bigger digital picture. As more information about the natural world becomes available, these maps become even more useful. Imagine if every car satellite-navigation system – or GPS-enabled mobile phone, for that matter – was able to send anonymous information about its location to a central traffic database. That database could build up an incredibly detailed real-time map of traffic congestion to make people aware of the traffic conditions ahead.[8] TeleAtlas is a service which already collects this data, which is used to ensure that maps are as accurate and up-to-date as possible. For example, if thousands of people are recorded driving across what appears to be a field on an existing map, it probably means that a new road has just been built (or there's a rave happening).

We are also developing new ways of collecting information about the natural world, which we're using to better understand the world around us. British scientists have developed a way to take some of the time-consuming tedium out of fieldwork. Instead of recording data out in the field, either with a computer or with paper, then having to download the data in the lab for processing, they've found a shortcut. The first part of it is a mobile phone application called EpiCollect that allows users to enter multiple bits of data into their phones, where they can be stored and then sent to a central database. Here, a second piece of software ingenuity – a web application – maps, visualises and analyses the data immediately as it comes in.[9] This approach is being used to track everything from the spread of disease to the distribution of rare species.

The use of mobile phones as data-collection tools is also giving rise to a new breed of researcher – the 'citizen scientist'. The New York City Cricket Crawl is an initiative to track the city's cricket and katydid populations, in the hope of finding signs of rare species such as the common true katydid (*Pterophylla camellifolia*). Volunteers are asked to record cricket sounds on their mobile phones, then send the recordings, as well as details of their locations, to a central number. The result is real-time census data and an interactive map of the insect population.[10]

This approach is even being used to track air and noise pollution in cities. American company Sensaris have developed a GPS-equipped sensor that can be strapped on your wrist like a watch, but which monitors air quality and noise pollution wherever you are. That information can then be transmitted via Bluetooth to your mobile phone, then to a central database, where it's quickly collated, analysed and made available to the public.[11]

But it's not enough to simply collect the data – you have to know what to do with it.

Putting it all together

Managing the flow of traffic or tracking the movements of a rare insect are easy, compared to the challenges faced by Australia's Water Information Research and Development Alliance (WIRADA), which is attempting to get a grip on the water resources and usage of an entire continent.

The project was born of frustration about the lack of data on Australia's water resources, according to Dr Rob Vertessy, Deputy Director (Water) at the Bureau of Meteorology. 'Myself and group of hydrologists were becoming increasingly frustrated that the business of hydrological modelling – predicting the state of water resources in the future – was forever constrained by access to good data,' says Vertessy. 'We'd seen a lot of technology trends that suggested to us that there is great potential here to modernise the way in which data is captured in the field, managed in databases and distributed over the internet.' The end result, after getting the government's support, was WIRADA – a collaboration between CSIRO and the Bureau of Meteorology that is charged with collecting and delivering comprehensive information about Australia's water resources.

That is a lot easier said than done. There are 240 different agencies around the country that collect data on water resources, so the project's first challenge is to collect all that data and convert it into a single format so it can be harmonised and put into a central database. The database is planned to go online in 2010, so that anyone with an interest in water management can easily access the data. 'The thesis I think we will be testing with our program is that with information at everyone's fingertips, it will inform a much more valuable and productive dialogue about water resources and support those that need to craft policy and management regimes to ensure that everyone has equitable access,' Vertessy says. For example, for people working within water agencies, this information can help with day-to-day management questions, such as whether to open or close a dam or a river gate. It can also be used for longer-term planning – for example, helping authorities to decide if, when and where to build a new dam or a desalination plant.

WIRADA's second challenge is to improve the breadth and quality of data on water resources. 'Hydrologic monitoring is a pretty crude business,' says Vertessy. 'The stock-in-trade measurement is measurement of flow volume past a point in the river – it's very hard to get that more accurate than plus or minus twenty per cent.' This is a particular problem when it comes to water trading – a market that is growing rapidly. In 2007–08, there was $1.8 billion worth of water trading going on, but Vertessy says the market won't really flourish until buyers and sellers have access to high-quality data. 'It is also particularly vital, when you've got a lot of trade going on, that you are properly quantifying the amount of water in the bank in the first instance, so that you're not buying and selling virtual water – you're buying and selling real water.'

This is where technological advances, such as acoustic Doppler metering, can help. These enable far more accurate measurement of the water in rivers under a range of conditions. This is particularly important when the trade in water is being made between states at opposite ends of a giant water system, such as occurs in the Murray-Darling Basin. Vertessy says about ninety per cent of water trade in Australia occurs around this system, but the vast distances the water travels demand accuracy in how the water is measured at points along the way, because some will be lost to evaporation or seepage. 'The longer the distance between where you're purchasing and delivering the water, the more critical it is that you're able to systematically track how much water there is in the system at various points and calculate all those losses.'

Bringing information together in this new world is known as *interoperability*. Not only are we storing more and more data about the natural world, but we are storing more information about the form of this data too. In the case of water and other natural resources, there is as much emphasis placed on the type and quality of the data as on the data itself.

So far we have described a one-way process, where things make their way from the natural world to the digital one. However, there is another major trend going on, in which data is moving in the other direction. As we start to place more and more services and information on the web, it is inevitable that things in the digital world will start to find their way back out.

Digital to natural

Meeting Mr or Mrs Right is a tricky business. Your eyes might meet across a crowded bar, but that's just the opening gambit of a very long and complex negotiation process that is riddled with questions. Are they single? Are they looking for love? Are you compatible? Never fear – with Bluetooth technology and mobile phone apps such as Proxidating, those questions are, thankfully, answered before you make a fool of yourself by chatting up the wrong person.

Proxidating – a contraction of 'proximity dating' – software uses your phone's location to look for any other Proxidaters in the area and checks out their profile to see if they are compatible with your preferences. If they are, it buzzes each of you with a picture of the other and a short text profile. You don't even have to be in a bar; the software is constantly on the prowl on your behalf and can 'introduce' you to your future partner on the street, in a supermarket – anywhere, as long as your phone is switched on. It's just one of many examples of how the digital world is intruding into the natural world.

The telecommunications analyst firm Juniper Research predicts that the mobile dating industry will be worth US$1.4 billon by 2013, up from $330 million in 2007.[12] But dating services are not the only ones exploiting the convergence between digital and natural. Mobile companies such as Loopt Inc. and Earthcomber help members use their phones to search for others nearby with similar interests, such as cycling or hiking.

Another technology that is bringing the digital sphere into our real-world experience is the Quick Response (QR) Code. It's a two-dimensional barcode that looks like a little black chequerboard, and was originally used to track components used in vehicle manufacturing. Have a look at the back of this book and you will see one. QR Codes are designed specifically to be read by a camera, which makes them perfect to be used with mobile phones.

When your point you phone at a QR Code and you have the right software,[13] you are taken directly to a website – in effect, the digital counterpart of the thing that you are pointing your phone at. Already you'll have seen QR Codes storing web addresses and URLs in magazines, on signs, buses, business cards, T-shirts – on just about any object that users might want to find information about. These codes have become a marker in the natural world for a digital location.

Advertisers have been quick to take up this and other new technologies. The advertising company Danoo, which specialises in digital screen advertising, has begun incorporating wireless technology into its ads, allowing viewers with Bluetooth-enabled devices to download information specific to the ad.

This is just the first step along the road to what is being called *augmented reality* – an overlaying of the real-world environment with digital or virtual information. The latest mobile phones come equipped with three very interesting pieces of hardware – a GPS receiver, a compass and a digital level. Using these together, you can work out exactly where you are standing, in which direction you are pointing and how you are holding the mobile phone. Couple this with a camera and some smart software, and you can start to overlay a view of the digital world.

One company implementing augmented reality is Layar.[14] Point your phone camera at something using this small piece of software, and it will start to put digital layers of information on top of the camera image, sourced from Flickr, Wikipedia, Google, Yelp, Twitter and many more. Look through your phone at a famous building and its history will appear. Look at a house for sale and you can view its past sale history, current rates and real-estate agent details. Forget street signs – the phone will point you in the direction of the nearest train station or 7-Eleven store. You may be even able to leave digital messages for people who will be at the same location in the future, or read them from those in the past. Get ready for digital graffiti.

Take away the phone and replace it with sunglasses – or even a contact lens[15] – and you get a view of the digital and natural worlds together – a head-up display for the world. But what would people look like in this augmented reality?

More human than human

It's a common complaint of parents today that the only way they know what's going on in the lives of their teenage offspring is by checking out their Facebook profile. Social-networking sites are becoming people's digital proxies, with detailed information about them, their movements, activities, interests, loves and hates – so much so that recalcitrant teens often do reveal more about themselves via their digital profiles than they do in the real world. These sites not only store information about you, which you may or may not care to share with others, but also about who you interact with.

So what will become the digital version of the human being? Social-networking sites such as

Facebook, MySpace, LinkedIn and Plaxo are only the beginning. There are applications like Twitter that allow you to broadcast what is on your mind, or like Dopplr that let you share your location with others, and still others like Flickr or Mozy on which you can store your memories and cherished data.

Virtual reality worlds such as Second Life take this even one step further, giving people the chance to create a virtual person – known as an avatar – as well as a personality. Second Life is a basic illustration of how we might all interact with the net in the future, and research is now being undertaken to look at how non-verbal communication works in this other world.

Social-networking sites and virtual worlds are two facets of the highly complex and rapidly evolving 'digital human'. Our mobile phones are another important facet. They are slowly picking up more and more information about us; in fact, in some ways they're becoming our way of interacting with the world. Your phone already knows where you are in a general sense, based on which cell site it is connected to. With the inclusion of GPS receivers, this information is being refined to within a few metres; phones are already starting to replace dedicated GPS devices for personal navigation.

Soon, your mobile phone will start to collect information about you that you might not even be aware of, such as your heart rate. If you're a fitness junkie, monitoring your heart rate during exercise helps you ensure you get the maximum cardiovascular workout. But if you have heart disease, or are at a high risk of developing it, real-time monitoring of your heart rate might save your life. Health diagnostics company Alive Technologies has developed a system that not only monitors heart rate in real-time but also transmits that information to a mobile phone via Bluetooth. The information can then be sent to a healthcare centre to allow doctors to monitor the health of their patients remotely.

This is particularly useful in monitoring arrhythmias – irregular heartbeats – such as a premature ventricular contraction (PVC), in which the heart occasionally beats abnormally.

PVCs are relatively common and can occur in otherwise healthy people of any age. For the most part, it's benign, but in people with existing heart disease, a PVC can trigger a more serious condition in which the heart beats too fast. But detecting PVCs can be tricky because the patient must be hooked up to an ECG machine when the PVC occurs, and this doesn't always happen. The Alive Heart Monitor can remain attached to a patient for twenty-four hours a day, monitoring and transmitting their heartbeat to a computer or a central collection point, thus making it much easier for a PVC to be detected and analysed.

Take this sort of technology to its logical conclusion, and we may see a future in which patients at high risk of experiencing heart attacks can permanently wear mobile monitoring devices. These would not only communicate their status to a central point for their doctor's information, but could also contact emergency services if a major interruption to the heartbeat were detected. A similar model might also apply to other chronic diseases, such as diabetes or high blood pressure, so that if someone experiences a sudden change in blood pressure or glucose levels, their doctor is quickly alerted and can check on their status remotely.

Your mobile phone will also become your wallet.[16] Going to the shops will be quite different. Rather than presenting your credit card for payment, you will instead show your mobile phone. A message will appear asking you to confirm the transaction – or perhaps to enter a PIN on your phone for verification. A survey by Forrester Research found that fifteen per cent of Japanese mobile phone users already make payments and purchases with their phones. Even more interesting is the fact that nearly one-fifth of Kenya's population of 38 million currently use mobile phones to transfer money to one another, buy goods and even pay school fees. Less time is wasted standing in bank queues or travelling to deliver cash, and time saved can be used to create opportunities to earn more. Some studies suggest that families who use the Kenyan M-PESA mobile money service have increased their net income by between five percent and thirty per cent.[17]

Don't forget all of the other stuff that you carry in your pocket or purse. Keys to the house? Use the phone. Swipe card for work? Use the phone. Phones may be used to travel on public transport, let you into the gym or any of the other things for which you currently carry cards, keys or tokens. Since phones can be deactivated from afar, it's a more secure form of identification than cards, which can be stolen. All these functions will be done using a technology called *near-field communication,* which allows communication over very short distances.

Eventually, you might not even need a mobile phone to connect with the digital world. Governments around the world are collecting swathes of biometric information about their citizens through photographs or fingerprints. Technologies that use unique characteristics of your body, such as retinal scanning, have been making leaps and bounds in recent years, to the point where computers in the future will actually be able to recognise you visually.

If you're starting to get a creepy feeling about the amount of information being collected about you, you're not alone. The convergence of our digital and natural personalities is causing some challenges, and there's a dark side to the new world. When such a massive amount of information about people is available in digital form, there is the potential for it to be stolen and abused.

People who already find the collection of personal information intrusive and threatening will be faced with an exponential increase in their concerns. Computer networks that, in the past, have remained separate will now be able to talk to one another, giving a depth of information to regulators that was previously unknown. Without serious policy action, our privacy laws may be left far behind.

Google Street View is one example of how privacy concerns and access to information are coming into conflict. Google's undertaking to create a digital 360-degree street-level view of cities around the world has raised privacy concerns because its images often include bystanders whose faces are easily identified. This has led to some potentially sticky situations, such as spouses caught lunching with someone other than their partners. In response, Google has begun blurring individual faces and car number plates, but it's still encountering resistance from the community in its attempts to photograph streets and house fronts.[18]

Others are justifiably concerned about information overload. With so much information at our fingertips, it will become harder to work out what is 'real' and what isn't, who is trustworthy and who isn't, and so on. Our digital personalities, moving through a digital world, will have to find ways to sort through large quantities of data continuously.

Assuming we can protect our privacy and find ways to cope with the inevitable information overload, however, the positive potential of these new technologies is enormous. This brave new world will bring you the information you need, when you need it and where you need it, enabling you to make more informed decisions and to interact with the digital and natural worlds in new and innovative ways.

A converged future

So what does this all mean in a world of limited resources? Virtual worlds and interactive advertising are fun and informative but, far more importantly, the convergence of digital and natural worlds will open up enormous opportunities for resource efficiency. The smart house, containing intelligent devices and connected to smart grids, will be able to monitor all of its resource flows, including water, waste and energy, and to manage these resources centrally with the equivalent of an Xbox or PlayStation in the living room. It will be attached to your security system and will know where you are. And if you want to save some money on your power bills, just run a program to increase your energy efficiency by switching off the lights when someone leaves a room, or limiting the kids' shower time.

There is big business in the measurement and management of these resources. Companies such as Tririga have now developed software that can manage resources on a much bigger scale, at the level of entire buildings or businesses – a process called *enterprise sustainability accounting.* Tririga's program first gathers information to build up a picture of a building's resources, included data on workplace assets, energy consumption and emissions data. It then identifies opportunities to improve energy efficiency, and analyses them from the perspectives of cost, financial return and societal value. It also manages the systems to ensure that a business continues to operate in the most sustainable manner possible. Mobile phone giant Nokia has already cut its energy consumption by seven per cent using this software.[19]

Convergence of digital and natural will also be a boon for planet Earth. Satellite mapping will enable improvements in natural-resource management and will underpin efforts in carbon and water accounting. In Liberia, individual trees are being barcoded so that timber exports can be carefully tracked from the moment a tree is logged to when it arrives at a port. The plan is not only to keep close tabs on Liberia's timber trade but also to combat illegal logging.

At the end of the sixth wave, we are likely to have a world in which the digital and natural have largely converged. Everything in the natural world will have a digital counterpart: you will be able to stroll through the web with your avatar, and information about people and places will be available through augmented reality on top of a planetary skin. In fact, it will be hard to tell where one ends and the other begins, and much of the digital world will be invisible to humans (as much of the natural world already is). For those looking to ride the sixth wave, opportunities will lie in finding something in the natural world that doesn't have a digital counterpart, giving it one and then working out how to use it.

Putting the natural and digital worlds together will provide increasing opportunities to synthesise data sets, thus creating value that didn't exist previously. This will also create opportunities for new businesses, which will exploit these gains in resource efficiency.

But with everything online and converged, what will be the shape of society? If we put together the big ideas of waste reduction and the convergence of the digital and natural worlds, we come up with two competing trends. On the one hand, physical resources will be used close to their original sources, in order to minimise waste and inefficiencies. On the other hand, information will no longer be bound by borders, and will be able to travel or be accessed from anywhere in the world.

How will we resolve these two competing trends? This leads us to the fourth big idea of the sixth wave: anyone dealing in 'atoms', or products and consumables, should think local, while anyone dealing in 'bits', or information, should think global.

Chapter 9

Atoms are local, bits are global

Dig! Dig! Dig! And your muscles will grow big
Keep on pushing the spade
Don't mind the worms
Just ignore their squirms
And when your back aches laugh with glee
And keep on diggin'
Till we give our foes a Wiggin'
Dig! Dig! Dig! to Victory

So it was that the British government encouraged its citizens to do their bit for the war effort as World War II raged around the globe, by creating vegetable gardens in every spare bit of land. The 'Dig for Victory!' campaign was born out of concerns over food security, as merchant ships importing food to Britain – much of it from the United States and Canada – became targets for the German navy.

At the time, Britain imported over 55 million tonnes of food a year, so there was an urgent need to find other ways to feed the population. As rationing took hold, private gardens, public parks, lawns and even sports fields were tilled and planted with crops, or turned into pens for chickens, rabbits and pigs. The campaign was so successful that by the end of the war food imports had been halved, while the amount of British land used for food production had increased by eighty per cent.[1]

There is a popular expression that 'civilisation is three meals away from anarchy', reflecting the fact that much of our food is transported to us over long distances, rather than being grown or farmed close to where it is eventually consumed. World War II may be long ended, but a new force is now driving a push for more localised food production. As fuel prices rise and peak oil looms, transportation costs have replaced enemy submarines as the motivator for change. Unfortunately, prime agricultural land is rarely collocated with the towns and cities that consume its produce, so produce has to be transported by planes, trains and trucks to where it is needed.

The EcoCity Farm is one solution to this dilemma. It is an *aquaponic* food production system – a system of cultivating plants and aquatic animals together in a closed-loop fashion. It's specifically designed for use in urban areas where land is a scarce resource. EcoCity Farms are the brainchild of third-generation farmer Hogan Gleeson and sustainable agriculturist Andrew Bodlovich, and was conceived more than fifteen years ago. 'We knew that peak oil would arrive and problems with climate change would arrive and we could see that at some point we'd need a local food solution,' says Bodlovich.

The challenge of small-scale food production in urban areas is to produce food that people can live on, and that means more than just vegetables. 'When you start looking at protein production in urban farming systems, fish are an ideal species because you can grow fish in a confined space without any odours or noises or any of the things that might bother urban people,' Bodlovich says.

The farm consists of a tank in which fish are farmed, a hydroponic system for growing plants, and a waste-processing module to convert waste into nutrients. It's a self-sustaining process. Aquatic species such as fish or crustaceans are farmed in the tank, while the plants are grown in nutrient-rich water, which is produced using waste from the aquatic species. The plants in turn filter the excess nutrients from the water, making it safe for use in the tank again. The system is almost a perfect closed-loop for water, such that ninety-five per cent of the water is purified by the plants and recycled back into the fish tank. Just five per cent of the water is lost to evaporation from the plants, and this is topped up automatically each day.

The various components of the EcoCity Farm are stacked one on top of another, minimising the use of space and allowing it to produce up to twelve times more food than other similarly sized systems. But most importantly, its small size means it can be set up close to where the end products are consumed.

'What we're doing in the first stage is the basis of an urban food production system that can be added to in the future,' says Bodlovich. 'What we can produce at the moment is the sort of crops you would typically produce in a greenhouse; for example, salad greens, Asian greens, strawberries, herbs, all of those sorts of things.' The next step would be large fruit and vegetables such as onions, potatoes and pumpkins, and perishable foods such as tomatoes that suffer from being transported over long distances.

It cuts out the middle man – transportation – saving money and time, and minimising waste. It's the sort of innovation we will see a lot more of in the sixth wave. In a world where waste is a source of opportunity, some businesses will exploit the efficiency of selling resources as close to the point of production as possible. But the shift isn't totally black and white: the move towards localisation for production-based business will raise the question of when to centralise and when to decentralise. At what point do economies of scale balance the cost of distribution?

And in that same world, where everything we use tends toward becoming a service, other businesses will exploit the global nature of information exchange. The internet will continue to remove barriers of distance, and electronic commerce will become increasingly globalised.

These two trends lead us to the fourth big idea: in the sixth wave, the size and distribution of a company's market will be decided by what it is selling. If it's in the business of selling atoms – consumables – then it will tend towards thinking locally. If it's in the business of selling bits – information – then its market is limited only by the reach of its communication media, such as the internet.

Once again, we see different approaches emerging in the biosphere and the technosphere. When we talk about atoms, we're largely focusing on things in the biosphere, where things we consume will tend to be sourced locally as the costs of transportation increase. In the technosphere, however, the cost of transporting information is low, and so services will be provided to any part of the globe.

In short – atoms are local, bits are global.

Atoms are local

How would you survive if you were stuck in the wilderness somewhere, far from supermarkets and corner stores? Alisa Smith and J.B. MacKinnon found themselves wondering this very thing while on holiday at their summer cottage, far out in the Canadian wilds. Instead of panicking and driving to the nearest town, they decided to see what nature was offering. They found trout, mushrooms, apples and rosehips – the foundation for a memorable feast that became the inspiration for a global food movement, the 100-Mile Diet.

After returning from their sojourn, Smith and MacKinnon decided that for one year, they were only going to eat produce that was sourced from within 100 miles (160 kilometres) of their home in Vancouver. By their own admission, it wasn't easy to begin with, particularly when it came to staples such as wheat. But as their knowledge of local food sources grew, so did their menu, until the time came when they were easily able to maintain their 100-mile resolve even when travelling to New York and Mexico's Yucatan Peninsula. Their journey has inspired many others around the world to undertake the same pledge to eat only locally produced food. The term *food miles* has even entered popular vocabulary, reflecting the distance that a product has to travel from pasture to plate.

The 100-Mile Diet is not a fad – quite the opposite, in fact. It may just be the way of the future. The Food and Agricultural Organisation suggests that in 2008–09, nearly one-sixth of the world's population was undernourished, and with climate change on the horizon the situation can only get worse.[2] Based on figures from the International Food Policy Research Institute, global demand for food over the next fifty years will exceed the historical food demand over the last 500 years.[3] Unfortunately, this will coincide with a major shift in weather patterns due to climate change, which will likely see dramatic changes in rainfall across most of the world's most important food-producing regions. We will also start to experience a shortage of fertilisers, as inputs to the farming process such as phosphate become harder to find. Agriculture is a large source of carbon dioxide, so it will be subject to increasingly powerful institutional forces like the pricing of carbon.

In short, we need to produce more food in the next fifty years than we have since the Renaissance, and we have fewer resources to do it with. That means that in the next half century we need to grow more grain than has been produced since the time of Leonardo da Vinci, harvest more fish than has been eaten since Christopher Columbus and extract more milk from all the reluctant cows than has been extracted since Queen Elizabeth I.

The solution is not just to produce more food, but to find ways of being more efficient in how we produce it. This means eliminating waste not only in food production, but in food consumption as well. Some of this wasteful consumption is in the inefficiency of the process itself, but a lot is lost just through transporting things around.

What are you really eating when you are eating food? A lot of the nutrients in food were once fertiliser. According to the International Fertiliser Industry Association, over 280 million tonnes of ammonium, 210 million tonnes of urea and similar amounts of ammonium nitrate, phosphates and potash are transported every year – a billion tonnes of fertiliser in total. If, as is currently estimated, this fertiliser is used to grow produce that feeds half the people on the Earth, each person requires roughly a kilogram of fertiliser every day to make the food they consume. Fertiliser is also shipped all over the planet, producing 37 million tonnes of carbon dioxide in the process.

And this is just the raw material. Once this fertiliser is converted into food, the food is then shipped elsewhere for sorting and processing. Packaging materials are also transported from all over the world to be wrapped around the food, and then the end product – your sorted, cleaned and shrink-wrapped pumpkin, for example – is shipped to you, the consumer. But even that isn't the end of the story. There's the waste to deal with – the discarded packaging and the leftover food. This has to be transported to a processing plant and either composted, put into landfill or, in some cases, flushed out to sea.

It's not only food that we are constantly transporting around the world. Just about every product we use in our day-to-day lives has been flown, trucked or driven long distances to get to us, and this comes at considerable cost.

All these atoms are quite heavy – and pricey – to transport. The price of transporting a kilogram of material can range from a fraction of a cent to tens of dollars, depending on the method used, and all of it requires more resources. This in turn generates waste products such as carbon dioxide. In a world where emissions from transportation are around fourteen per cent of total global emissions (and up to a quarter of the emissions of some developed countries), it stands to reason that if we are to reduce our carbon footprint, a very good place to start is by reducing the amount of goods we transport.

For all these reasons, the incentive is building for industry to rethink traditional supply chains and to take a more local view – this is particularly important for things we consume in the biosphere. Institutional changes such as carbon cap and trade schemes mean that the true cost of transporting goods is finally being realised and incorporated into the cost of a product. This not only hits businesses but also consumers.

Consumers, in fact, are beginning to demand to know these total costs. The past decade has seen the rise of movements such as Slow Food and numerous 'buy local' campaigns. These efforts are being organised and advocated not only by environmental groups but also by trade organisations keen to retain local jobs. Environmentally minded consumers are voting with their wallets, and the trend towards local foods is creating a marketing edge for some providers, as well as financial savings. But food production isn't the only industry looking local – the entire product marketplace is shifting its focus in this direction.

Tailor-made

If you own a relatively new car, the loss of a particular engine part, while inconvenient, is not a huge crisis. At worst, your car might be out of action for a few days while your mechanic finds a replacement. But if you drive a much older model or an antique car, a broken part can be disastrous. If it's irreplaceable, your beloved automobile may well have done its dash.

But imagine if you had a system that could scan your broken part in high resolution, then rebuild it from scratch, layer by layer, copying the original down to the finest detail, and do it all in your garage. It's not as far-fetched as it sounds.

Three-dimensional printers and scanners are already starting to bring the factory to your home. Picture this: It's late in the evening and you've finally finished an important report that you have to present first thing in the morning. You hit 'print', and nothing happens – except for an ominous clunking noise. Deep inside the printer, a small but vital piece of plastic has snapped, and it couldn't have happened at a more inconvenient time. Repair shops are closed, printing shops are closed, and you won't have time to print the report before the presentation.

What you need is a RepRap – a self-replicating rapid prototyper. This is a three-dimensional printer that can 'reprint' objects from a computer model. RepRap works by building up layer after layer of plastic until it has created a three-dimensional object.

So far, the RepRap, which has been developed by around 100 people from all over the world as an open-source project, has produced everything from coat hooks to children's sandals, as well as replacement parts for other RepRaps.[4] With such a device close to hand, you could simply reprint the broken part without leaving your house. While RepRap is so far limited to producing plastic parts, it is at the vanguard of the local-production movement. In the future, consumers may use it to create toys, small replacement parts for devices at home or even whole new projects ordered online. Even better, being open-source, RepRap also comes with the code to replicate itself.

The home 3D printing industry may be in its infancy, but already companies are using commercial 3D printers such as Objet for things like rapid prototyping and one-off manufacture. Objet's technology uses inkjet technology, except that rather than printing ink, the technology prints a layer of polymer a fraction of a millimetre thick, which is then cured with ultraviolet light. The printing tray is lowered and another layer is printed. Layer by layer, a solid object emerges. It might be the prototype of an artificial hip joint, the latest Doctor Who toy collectible or a detailed scale model of an office building.[5]

On a larger scale, new technologies in flexible production are allowing small batches of different products to be made by the same factory – the idea being that one factory can make a whole variety of products for a local market as required, rather than a local market having to import each of those products from specialised factories all over the world. Just as we can take information from the natural world and put it into the digital one, now we are able to take three-dimensional objects from the digital world and build them in reality.

Germany's Fraunhofer Institute for Laser Technology has pioneered a process known as *selective laser melting,* which allows metal and ceramic machine components to be produced directly from a computer model. The process starts with a powder that is deposited in a thin layer inside a process chamber. Following the computer model, the powder is melted with a laser beam to create the first layer of the component. Then a new layer of powder is deposited, melted and fused to the layer below it. This is repeated until the part is finished.

The beauty of this process is not only that it can make single batches of products on demand, but also that it enables products of nearly unlimited complexity to be created, as the parts are built up in layers. The Fraunhofer Institute sees this flexibility of production as a way to add more value to the manufacturing process, and thus to compete with manufacturing powerhouses on the world stage.

A world in which products can be tailored to exact specifications and produced locally has many advantages. Companies can outsource their production facilities to local multipurpose factories, reducing lead times for products and limiting unnecessary stockpiles and transportation costs.

The biggest advantage of these flexible factories and 3D printers is that they enable fast, local production of goods, which previously would have taken many weeks or months to produce and might have required the services of a specialised manufacturer on the other side of the world. In the new scenario, localised production saves time and money. But there's a third advantage to thinking local, and that is cutting down on waste.

Distributed infrastructure

There's nothing worse at the start of a long-anticipated holiday to arrive at your destination and find that your baggage has taken a detour. Baggage lost in transit costs the global airline industry an estimated US$1.7 billion each year, but this pales into insignificance alongside the transmission losses experienced by the US power grid. A study found 9.5 per cent of all power generated in the United States is lost during transmission and distribution.[6] Transmission losses are of particular concern when it comes to renewable energy. A solar power plant or wind farm is rarely going to be located close to civilisation – they're more often built in deserts or barren, windswept regions that few people would wish to call home.

One solution is high-voltage direct-current technology. This uses direct-current instead of alternating-current cables, which can carry more power, with less leakage and greater stability, and all at a lower cost. High-voltage direct-current cables already connect Britain with France's nuclear power, and enable the exchange of renewable energy between Denmark and Norway.

Even bigger plans are being mooted for a green 'super-grid' of high-voltage direct-current cables, connecting Europe to sources of renewable energy further afield, such as geothermal and hydroelectric power in Iceland and solar power in North Africa. It would enable the transmission of massive amounts of electricity across Europe to meet demand with less transmission loss. Such an enormous network would also be able to absorb some of the variability of renewable energy sources, meaning that when the sun sets over the solar power stations on one side of the continent, the grid can draw from geothermal and wind power on the other side without any interruption to supply.

However, even technologies such as high-voltage direct-current cables are not completely lossless, which leads us to another possibility: why not get rid of the distribution problem in the first place? *Distributed infrastructure* is infrastructure that doesn't come from a single source but is spread throughout a community. In the energy space, this might be through solar cells on roofs, or local sewerage treatment (and recycling) plants. In the past, economies of scale have made these solutions less attractive than large centralised plants, but as the world focuses on the efficiency of its natural resources, could it be that these technologies start to look more attractive?

Distributed energy is already used in cities around the world, although it is still in its infancy. Roofs are sprouting photovoltaic panels and solar heaters, and even small-scale windmills are starting to appear in backyards. However, when solar panels come down in price and electric vehicles can store energy in their batteries, a whole new energy infrastructure will emerge. At the moment, for example, a house equipped with a photovoltaic array can use the energy it generates during the day to power essential services, and whatever excess energy it produces is fed back into the main grid. But perhaps a better way would be for that excess energy to be stored in the batteries of the electric car parked in the driveway. After the sun goes down, the car then feeds that excess power back to the house, at least during times of peak load.

As solar technology improves, more houses and buildings will have the ability to generate their own energy. While solar panels are already being fitted on many homes thanks to generous government subsidies in areas such as California, they are generally fairly bulky and don't necessarily fit with the look of the house. However, newly developed thin and flexible solar cells are now being incorporated into the building products themselves. The US company SRS Energy's Solé Power Tile integrates a flexible solar cell into a conventional clay roof tile, so an entire roof can be built of solar panels without changing the look of the house. There are even solar panels so thin and flexible that they can be used as a coating on top of other materials such as corrugated iron. Even the old shed out the back may become a source of energy for the house.

Distributed energy isn't just about finding new ways of generating power; it's also about being more efficient with existing technologies. *Cogeneration* – also known as *combined heat and power* – is one approach becoming increasingly popular around the world. As the name suggests, combined heat and power systems produce both heat and electricity, usually by capturing the waste heat produced during electricity generation and then using it for space or water heating. This even extends to what are called *tri-generation systems,* which combine power, heat and cooling technologies. District heating is an extension of this, where heat created during the generation of electricity is used to warm buildings in the area surrounding the power-generation facility. Given the challenges in moving heat around, many of these systems only become efficient when the power systems are 'close to home'.

Distributed energy has already evolved to the point that some houses generate all their own power, capture all their own water and treat all their own sewage. These are totally self-sufficient and so are 'off the grid'. Tony Marmont is one of these so-called 'off-gridders'. His family farm in the United Kingdom is a monument to distributed energy. Heating is supplied by a combined heat and power system fuelled by propane, which is ninety-five per cent fuel-efficient, and a water-sourced heat pump. This system operates out of the nearby lake, taking advantage of the fact that the lake's water maintains a reasonably stable temperature. It transfers the small amount of heat from the lake into a refrigerant with a very low temperature of vaporisation, and then a heat pump extracts this heat and stores it.

The house also has a wood-fired stove for comfort and aesthetics, and water is heated partly by the heat pump and partly by solar collectors on the roof. Electricity is supplied by two wind turbines, two photovoltaic arrays and two water turbines – one in the lake and one in a nearby stream. To store excess power for when the sun goes down or the wind drops off, Marmont's farm features a sophisticated system where excess power is used to drive a chemical process called *electrolysis,* which splits water into hydrogen and oxygen gas. These are stored in pressurised tanks; when needed, they are recombined in a process that generates electricity and hot water.

Is there a business behind all this? The deployment of solar panels, wind generators or micro-generators might mean that some buildings are net power exporters. We've already seen that some solar-power companies are appreciating the value of roof space and are renting it from consumers to install solar panels. In the not-too-distant future, we might start to see financing companies come up with new models to sell previously unused space in cities.

This sort of configuration is also quite attractive for a number of other reasons. Firstly, it makes the grid more balanced. Congestion in the current system can cause bottlenecks in parts of the grid, leading to power blackouts and disturbances – the US Department of Energy estimates this costs the nation's economy as much as US$180 billion each year. Town planners also like it because the infrastructure is very robust. Even if some parts of the system fail, other parts can take up the slack. While it is a significant change to the business model, some energy companies are realising that it's better to ride the wave than be left behind.

Other infrastructure is going local too. Those who remember the time before sewer mains – or those who still live in regional areas – will recall the joys of the septic tank. They did the job, but were often smelly, would require some unpleasant chemicals and would need to be emptied from time to time. No wonder mains sewerage was a blessing. Thankfully, technology has advanced considerably and now we have systems such as Biolytix's BioPod.

The BioPod is a household treatment system for all waste water, including sewage. It uses natural micro- and macro-organisms to break down and process the waste. The system consists of a 3000-litre polymer tank that contains layers of filter beds. These house the organisms, which include bacteria, earthworms and beetles. As the waste and water are pumped into the tank, the solid waste is converted into humus, which also filters the water, while the remaining water is carefully filtered to remove anything larger than eighty microns in size. This water is then pumped out for use in irrigation. The best thing about this system is that it regulates itself – the microbes adapt to the amount and type of waste matter.

Where will this decentralisation end? That isn't a simple question to answer, as there is always a trade-off between the efficiencies created with economies of scale and those created by cutting down on transporting all those atoms around. There's no question that global supply chains will still be critical to the world economy, but we'll see a move away from centralised production, supply and infrastructure as the norm, and institutional changes will shift the balance towards more local technologies.

Bits are global

Fancy some anchovy ice-cream? If you answer 'yes' to this, it's fair to say you're probably in a very small minority. And if you do happen to like anchovy ice-cream, it's probably also fair to say that there aren't many ice-cream manufacturers in your neighbourhood that cater to your particular, and some might say peculiar, taste. (No offence.) Because of high production and storage costs, there's little incentive for firms to cater for the single person who wants anchovy-flavoured ice-cream.

In traditional (local) markets, the majority of consumers want the same thing, and businesses respond by producing it; in the case of ice-cream, we're most likely talking chocolate and vanilla. But when you start to put t o g e t h e r all t h e anchovy-ice-cream-lovers in the world, a market does emerge. The more efficient the production, storage and distribution of ice-cream becomes – or the more flexible the local manufacturing processes become – the more viable it is for an ice-cream company to supply unusual flavours to those who have unusual tastes. This creates significant opportunities for firms to capture global markets that were, until recently, unviable at a local level.

In 2004 the editor-in-chief of *Wired* magazine, Chris Anderson, coined the term 'the long tail' to describe a strategy for businesses who sell a large number of unique items, each in relatively small quantities. Businesses, starting with advertising and media, have been exploiting this long tail to find global niche markets for their products.

This is the future for the technosphere. Because it costs almost nothing to move information around the world, businesses dealing in the technosphere – those that deal in *bits,* information and services – have the opportunity to access global niche markets from day one. Television is a good example of this. Most cities started with only a few television networks, which each cater for different segments of the market. Some are focused on news, some on sport, and others on sitcoms and drama. But the arrival of the internet brought new opportunities for television to be tailored to individual tastes with shows that would never see the light of day on mainstream networks. For example, Apple's iTunes sells episodes of old, obscure sitcoms, while YouTube has enabled advertising campaigns to spread virally. Internet-only television shows such as *Yacht Rock* have even developed their own cult followings.

Another example is internet radio. Restrictions on AM or FM bandwidth mean you might only be able to pick up a handful of radio stations in your area, which between them might cater for the young and funky, the golden oldies, the middle-aged, the news-hound and the classical buff. But if you want true variety, get a device like the Roku SoundBridge, which picks up internet radio stations around the world. Fancy some chillout electronica, island metal reggae or religious funk? There are now over 12,000 radio stations around the world broadcasting over the internet, and because there is little cost associated with internet radio, broadcasters can afford to create a station that caters to the musical equivalent of the aficionado of anchovy ice-cream.

Another feature of the long tail phenomenon is the rise of the 'prosumer' – a merging of 'producer' and 'consumer'. Coined by the futurist Alvin Toffler in his 1980 book *The Third Wave*, the term describes a new breed of consumer more actively involved in the production as well as consumption of goods. Catering to this emerging class of individuals is the Design To Order system, invented by Japanese entrepreneur Kohei Nishiyama. Design To Order takes normal product development and turns it on its head – users are invited to submit their ideas online for a new product, or to place an order for a product that someone else has come up with. Users of the site are given access to tools to design their products and discussion forums to pitch their ideas. Once a certain number of orders are placed, a manufacturer is found to make the product and the original designer gets a cut.

For example, say you find the chair at your local dental clinic quite uncomfortable, and you come up with a brilliant idea for a better chair. 'So you have a new idea of how the chair should be at the dentist; you put it up [on the website] and amazingly you find that there are a thousand patients who need the same kind of chair and would prefer to go to a dentist who has this kind of chair,' says Nishiyama.

A company that has the capacity to manufacture this particular type of chair is therefore able to see a clear market for this new product. They can approach dental clinics to buy the chair, securing orders for the product before it has even left the assembly line. Dental patients who have given the thumbs up to the new design through the website also act as marketing agents for the chair by convincing their local clinic to invest in the improved design. And best of all, the chair pushers are rewarded for their efforts with a small commission on the sale of each chair – similar to the affiliate fees offered by other online ventures such as Amazon and eBay.

What Design To Order does is tap in to the long tail and convince users to get involved and share their preferences and ideas. Rather than wishing for a product, such as an improved dentist's chair, you can now imagine it and see how many other people want one too.

There are benefits for all involved, says Nishiyama. 'Manufacturing will have less risk of producing extra amounts of inventory, user will have less risk of purchasing something they don't want to have,' he says. 'It reduces the amount of materials or natural resources or energy that we spend so we are producing just the optimal amount that we need to consume.' It also cuts out a lot of 'middle men'. 'We are moving the artefact to the final destination, directly from factory to consumer without passing through the retail stores or any other warehouse that was built for inventory or marketing purposes.'

Nishiyama started commercialising the system in 1997 through his company, Cuusoo. Design To Order has resulted in a range of new designs and products, from improved laptop bags to funky light fittings. The lifestyle company Muji uses the system to source ideas, with Cuusoo products often ranking in its top three bestsellers. LEGO is even using Design To Order to create new designs for its popular blocks.

The phenomenon of the long tail is not linked only to the weird and the wonderful, nor does it solely cater for people with exquisite tastes. In many areas, technologies such as social media, low-cost services and the increasingly global outlook of society are driving the long tail to become even longer.

Take international travel, for example. Over the past ten years many low-cost carriers have emerged, offering point-to-point travel and decreased transaction costs through online booking services. Search engines have made more obscure locations more available, and social networks have allowed more word-of-mouth recommendations to spread. Consumers are also able to share tips and, as a result, are demanding more authentic experiences. All of this has increased the diversity of locations that people are travelling to. In 1998 thirty-six per cent of British tourists travelled to the same top fifty destinations – the 'head' of the curve. By 2008 this had dropped to twenty-six per cent, with almost three-quarters

travelling to more obscure destinations
– the 'tail'.[7]

Some have called this the
'democratisation' of both information
and services. As more choices and more
information become available, consumers
will start to exercise the more varied
options available to them. Companies
that do not accept and cater for this
increased consumer power but continue
to offer the same one-size-fits-all
product or service will see their market
share eroded.

But there is a trap in being too
global: try to please everyone, and
you'll end up pleasing no one. While
those dealing in bits can afford to think
at a global level, and indeed would be
foolish not to, this doesn't necessarily
mean that their product or service will
be wanted in all countries and markets.
They still have to think local.

Glocalisation

What do you get when you mix
global business with local preferences?
Glocalisation. It doesn't quite roll off the
tongue, and is similarly complex to

master, but global businesses have to come to grips with it if they are to succeed. The idea behind *glocalisation* – a mashing of *globalisation* and *localisation* – is that a global product or service is more likely to triumph if it is tailored to local preferences, practices and culture.

Craigslist began life in 1995 as a very local concept – an email list of events in the San Francisco area, sent mostly to friends of the founder, Craig Newmark. After moving to become a web-based service, Craigslist mushroomed into a global phenomenon. There are now Craigslist sites in more than 700 distinct locations in seventy countries. More than 50 million people use the service in the United States alone,[8] looking for love, housing, bargains, jobs, study courses, entertainment, tradesmen, debate, sellers, buyers ... the (Craigs) list is endless.

While Craigslist is a truly global dealer in bits, it is also extremely localised. It wouldn't have been anywhere near as successful, and would probably have failed early on, if there was just one Craigslist. Imagine the frustration of someone looking for housing in Madrid, only to be offered a list of houses in Tokyo, London and the Virgin Islands; or the irritation of looking for employment in real estate and finding that your dream job is on the other side of the world. What makes Craigslist special is the fact that it is targeted at a specific location.

As GPS technologies are incorporated into many devices, the opportunities for these global but local services emerge. Many of the applications for the Apple iPhone have a local component; an app might automatically find the nearest restaurants in your vicinity and choose one at random, or direct you to the nearest train station. These global offerings of local services have an edge that makes it hard for more traditional businesses to compete with them.

Firms and technologies such as Kohei Nishiyama's also open a whole set of opportunities for personalisation – businesses that offer tailored services from a global platform. As the digital and natural worlds converge, more and more information about our personal preferences is becoming available, and firms will start to use this information as a source of competitive advantage. United with small-batch or single-item manufacturing, ordering products that are completely personalised to your tastes may become ever easier.

Personalised services could also hold the solution to privacy concerns. Giving a customer the option of releasing personal information in exchange for a more tailored service allows them to choose how much privacy they wish to retain. This already happens with some schemes, such as frequent flyer programs, where people allow themselves and their travel habits to be monitored by an airline in exchange for points, improved service and a more tailored experience – their preference for an aisle seat, for example. Personalisation may soon appear in everything from healthcare to food production – just imagine a breakfast cereal tailored not just to your personal taste but to your genetic disposition.

Large international firms are also seeing the opportunities in global business with a local flavour. Global products firm Johnson & Johnson maintains the integrity of its global brands but also allows its marketers a significant amount of flexibility to tailor products to consumer preferences in their local markets. For example, the company sells four-packs of Band-Aids in developing countries, rather than the larger box that's available in the developed world, after seeing how local distributors made up their own smaller packages of Band-Aids to sell.[9]

Others are even taking advantage of technologies to source local talent and ensure that products and services stay relevant. txteagle is a form of 'mobile crowdsourcing' that taps into both the need for local knowledge and the abundance of human resources in the developing world. There are now two billion people who own a mobile phone but have an income of less than US$5 a day; txteagle provides them with the opportunity to earn additional income by doing simple tasks using the text functionality of their mobile phones – from straightforward image-tagging to more complex transcription or translation tasks.

txteagle recognises that for certain tasks, there is no substitute for human beings. Take transcription, for example – the global market for transcription services is currently worth over US$12 billion per annum and is growing rapidly. txteagle pays mobile phone users to transcribe passages of text, and employs multiple users to check the quality of each transcription. The translation market itself is also significant – txteagle is particularly useful for translating into local languages or dialects, and particularly lucrative in the expensive medical translation industry. The same approach can be used for everything from citizen journalism to increasing the relevancy of search-engine results.[10]

The opportunities available when 'going local' have implications for large multinational companies that have traditionally been headquartered in a single country – think of the relationship between Honda and Japan, or Caterpillar and the United States. As production and marketing become more local, and as supply chains and the search for talent span global boundaries, this relationship may start to change. Minhir Desai from the Harvard Business School describes this as the 'decentering of the global firm' – a decoupling of the relationship between large firms and their home countries.[11]

Straddling global and local is a difficult task and requires a unique set of skills that will be highly sought after in the sixth wave. Those who do this – 'glocals' – may even comprise a new social class: people who are closely connected to their local communities, yet who work on an international stage. They choose where they want to live without giving up their global connections.

A local, global future

When to focus locally and when to think globally? When to centralise and when to decentralise? Sixth-wave thinking requires you to keep both in mind.

Atoms are local. If you run a company that deals in consumables, think about strategies to manage your resources locally as often as you can.

This can range from distributed technologies such as decentralised power generation, to local sourcing, recycling and distribution. Your competitive advantage will come when full life-cycle costs are incorporated into your product; you can then establish where, on the scale of centralised to decentralised, the cost of resources balances the cost of distribution.

Bits are global. If you run a company that deals in information and services, think globally from the beginning, as you are likely to be not bound by resources, distance or location. You might find strength in niche markets or the long tail, or through the local delivery of your global service.

These distinctions aren't fixed, of course, and like individuals, businesses can exist in both domains. But when thinking about your business model, ponder this: are you 'born global' or do you have 'local roots'? Both can be a source of competitive advantage.

The four big ideas of the sixth wave we've seen so far – competitiveness of waste, the shift to services, the convergence of the digital and natural worlds, and the trend towards glocalisation – are all, at their heart, about resource efficiency. The more efficient we are with resources, the greater the likelihood that our species can co-exist in harmony with the rest of the natural world. But perhaps the biggest sixth-wave idea might just come directly from the natural world we are striving to preserve. In the next chapter, we'll look at how nature may hold the greatest insights of all.

Chapter 10

If in doubt, look to nature

As sea urchins go, the Antarctic sea urchin (*Sterechinus neumayeri*) is fairly unremarkable. It's a small red creature that resembles a pincushion with a penchant for dressing itself up in bits of shell, debris and any handy stinging sea polyps that happen to be lying around. However, *Sterechinus neumayeri* does have one impressive distinguishing feature – it is one of the most energy-efficient creatures on the planet, largely due to its days of living in what can only be classified as a freezer.[1] All animals use, on average, about thirty per cent of their available energy in metabolic processes – in particular, to build new proteins – but not this little critter. The embryos and larvae of this remarkable urchin use twenty-five times less energy in performing this vital function. Such incredible energy efficiency enables the urchin to survive and thrive in an extremely harsh environment where resources are scarce.

It's a century and a half since Charles Darwin shook the world with his theory of evolution, which proposed that the organisms best suited to survive in their particular circumstances had the greatest chance of passing their genes to the next generation. But 'survival of the fittest' isn't just about two animals fighting tooth and claw – it can be about how widely a tree disperses its seeds, how sensitive an animal is to predators, how skilful a bird is in laying her eggs somewhere safe, or how energy-efficient an urchin is in an environment where energy is at a premium. Natural processes balance competition, cooperation and coordination, on the one hand, and elimination of waste, on the other. Individuals fill the niches opened or left open by others who are less efficient.

With billions of years of evolution under her belt, it's not surprising that Mother Nature has developed a few tricks in the resource-efficiency game. The natural world is balanced, efficient, resilient, responsive and constantly evolving, and it can provide abundant inspiration and information for a species that is yet to master this balancing act. With this in mind, our final piece of advice in a world of limited resources probably comes as no surprise: if in doubt, look to nature.

Copying life

Had George de Mestral not been into hunting, this electrical engineer from Switzerland might well have led an unremarkable life and remained in obscurity. But while on a hunting trip with his dog in the Swiss Alps in 1941, he brushed past a patch of burdock thistle, which deposited several burrs on his clothing and in his dog's fur. After removing them, no doubt with some difficulty, he examined them under a microscope. He noticed a pattern of tiny hooks on the seeds' surface that enabled them to cling tenaciously, but not permanently, to anything furry. The discovery famously led to the invention of Velcro and assured de Mestral of a place in the history books.

Humans have always looked to their natural surroundings for inspiration. At first, that inspiration might have been born out of necessity – early knives created to defend against slashing claws – but since then we have become more aware of the possibilities in the natural world around us. From snowshoes that mimicked the feet of the snowshoe hare to Leonardo da Vinci's early sketches of airplanes that drew heavily on a bird's wing, we have looked at nature's designs and copied what we could. This is the field of *biomimicry* – literally, 'imitating life' or, as biomimicry expert Janine Benyus puts it, 'innovation inspired by nature'.

'There really is no better model, because life has had a very long time to do research and development,' says Benyus, the co-founder of the Biomimicry Guild and a natural-sciences author. According to Benyus, there are nine laws that underpin design in nature.[2] These range from the obvious (that nature runs on sunlight) to the more profound (that nature curbs excesses from within). If you reward cooperation, use only the energy you need, recycle everything and demand local expertise, you are starting to get close to the way that nature designs things.

Nature is also an ideal model of sustainability. '[Living organisms] have learned how to do what they do without toxins, at life-friendly temperatures, and perhaps even more importantly, they have learned to meet their own needs while enhancing the place that's going to take care of their offspring,' says Benyus. 'It's life creating conditions conducive to life.'

There are several key principles that describe how life remains sustainable. The first is what Benyus describes as 'benign manufacturing'. Unfortunately, almost all of our manufacturing processes rely on heat, and in that sense they have evolved very little since the dawn of civilisation. 'It's the first technology and the one we're in thrall to,' Benyus says. But the tough antlers of the caribou or the hard shell of a mollusc don't require heat to be created – they use chemistry. This has given rise to the field of bio-inspired green chemistry, which looks to nature for manufacturing solutions.

The second important principle we could emulate in our quest for sustainability is local procurement. 'That's really important, in the sense that it's not just shopping locally, it's also getting into very short, immediate feedback loops with your local environment,' Benyus says. 'So an organism really is aware of the limits of the place and the opportunities of the place, and they make the most of the habitat that way and are constantly working to fit themselves in within the parameters of those limits and opportunities.'

Another important principle is that of resilience, which is complex enough that it has given rise to an entirely new field of research – *resilience science.* 'This is the study of why, when a catastrophic event happens, like a fire or storm or tsunami, two forests will have the same fire and yet one will completely fall apart and become a shrubby grass field and the other will get burned up and come back as a forest.' Nine different properties have been identified that make a difference between collapse and survival, says Benyus. These include how interconnected a system is, how much information flow there is and how diverse it is.

While there are plenty of examples of innovators taking a leaf out of nature's design book, we don't do it nearly as often as you might think. A study by researchers at the University of Bath's Centre for Biomimetic and Natural Technologies compared the inventive principles underlying human solutions to problems with the inventive principles most often observed in nature.[3] It found there was only a twelve per cent overlap between technological solutions and biological ones. Human technology tended to solve problems by manipulating energy, the researchers reported, while biology more often relied on information and structure. It suggests there is still a lot we can learn from the biological world, and the field of biomimicry is indeed breaking some extraordinary new ground.

Just ask Jaws

Few predators on Earth are as feared as the shark, and for good reason. It is a supremely engineered killing machine, bristling with innovations that make it silent, swift and deadly. One of those innovations is its skin, which is covered with tiny structures called *denticles* – literally, 'small teeth'. These specially adapted scales point backwards, so that if you rub your hand down a shark's body, it feels smooth, but when you rub against the direction of the denticles it feels as rough as sandpaper. As the shark moves through the water, each denticle creates a tiny vortex in the water, which reduces drag and turbulence, making the shark not only faster but also quieter.

When swimsuit manufacturer Speedo was researching how to improve its swimwear fabric for competitive swimmers, it looked to the shark for inspiration. The result was Fastskin. This unique fabric is covered in ridges that copy the size and shape of denticles, in order to shave those critical few milliseconds off a swimmer's time.[4]

Shark skin also has another interesting property – it doesn't collect barnacles, which would increase drag and slow it down. For the same reason barnacles and other marine organisms such as algae are the scourge of the maritime industry. Because the denticles are able to move and flex individually, it's much harder for barnacles to get a grip on the shark. Researchers from the University of Applied Sciences in Bremen, Germany, were inspired to create an artificial sharkskin of elastic silicone that can be applied to a ship's hull; their invention has reduced fouling by sixty-seven per cent. The denticles make it very hard for barnacles and algae to stay attached – a speed of just five knots is enough to knock them off, effectively enabling the ship to clean itself. Given that fifteen per cent of all drag is created by barnacles, mussels and other sea creatures, this translates into huge savings for maritime industries.[5]

It doesn't stop there. Sharks (and fish in general) are very efficient swimmers, inspiring researchers at Boston's Franklin W. Olin College of

Engineering and Boston Engineering to create GhostSwimmer. This autonomous underwater vehicle looks nothing like a traditional submarine and every bit like the bluefin tuna that influenced its design. Instead of a propeller, GhostSwimmer is propelled through the water by a tail fin, which is up to twice as efficient. In fact, it can operate autonomously, calculating how to swim with maximum efficiency.[6] This design can not only go three times as far on the same amount of fuel, but it is quieter, more streamlined and may even be able to travel faster.

GhostSwimmer's mimicry of a bluefin tuna even extends to its internal structure. Instead of a pulley and cable system to operate the tail fin, GhostSwimmer has a spine with vertebrae, and synthetic muscles – electroactive polymers that change shape when an electric impulse is applied. These enable the entire body to flex and bend in much the same way as a tuna would. Other teams are looking at the hydrodynamic properties of dolphin fins to improve the efficiency of aircraft.[7]

The hydrodynamic features of marine life can also be used in reverse – instead of propelling something through the water with greater energy efficiency, these features are being adopted to improve the efficiency with which the movement of water is converted into energy. bioSTREAM is a tidal power conversion system that consists of a tail fin attached to a pole fixed to the seabed.[8] The fin, which can rotate freely with the direction of water flow, mimics the motion and characteristics of the tails of such impressive swimmers as the tuna, shark and mackerel. As water flows past, the energy of the flow moves the fin, and that movement in turn drives an electrical generator. The shape of the tail fin means it captures as much energy as possible, with little lost to turbulence or drag.

There is still so much more we can learn from the shark. For example, their sense of smell is incredibly acute, thanks to a nasal structure that enables an extraordinarily large number of smell receptors to be packed into it. Some sharks are able to detect blood in seawater at a concentration of just one part per million, while others have the ability to sniff out the volume equivalent of a golf ball in Loch Ness. Sharks are also acutely sensitive to electrical fields, being equipped with modified hair cells that respond to changes in electrical polarity.[9]

And if this is just the start of what we can learn when we take a closer look at the shark, what else might be out there? Sharks are just one small collection of species. What insights might the billions of other species on the planet hold?

A tour of nature unearths some astounding materials. For example, spider silk is one of those materials that is leaving human engineering in the dust. It is many times stronger than our best steel, yet it's incredibly lightweight, can stretch to 140 per cent of its length without breaking, and is able to maintain its strength at extreme temperatures. However, it has been a struggle to use it in an industrial application, as spiders aren't especially cooperative when you try to milk them of their precious silk, and we still don't know enough to be able to mass-produce it artificially.

But the more we turn to nature, the more we find. Professor David Kaplan from Tufts University, Massachusetts, has developed a way to incorporate artificial spider silk into a lightweight synthetic material known as a *hydrogel* to make it stronger. This makes it suitable in situations where weight is at a premium, such as in aircraft or when coating biologically friendly materials in medical procedures. Silk from spiders and silkworms has the added advantage of being biocompatible, which makes it particularly useful in medical applications. Kaplan and his colleagues have been developing an artificial bone matrix using dissolved and re-engineered silk proteins. This bone matrix can be used to rebuild damaged bone and even to regrow teeth.[10]

From a tiny spider to the giant redwood – trees are also astonishing the scientific world with their abilities, particularly when it comes to moving water around. A tree is able to draw huge amounts of water from the ground up into the highest branches and leaves, without a pump in sight. Their secret is the intricate network of tiny tubes called *xylem,* which pull water up through capillary action. Cornell University researcher Abraham Stroock is attempting to replicate this action using a hydrogel. His 'synthetic tree' is not just a curiosity – this ability to draw fluids through a compound will have applications from medicine to architecture. It could be used to create wound dressings that are able to remove fluid and deliver medication evenly throughout a wound, or to circulate heat around a building by moving water warmed by a solar collector.[11]

Naturally inspired

Sometimes biomimicry is serendipitous – an inquiring mind wonders how something in nature works, and that inspires them to invention, as was the case with de Mestral and velcro. However, it is beyond serendipity that biomimicry has delivered its greatest benefits. Solving a problem through biomimicry requires you to think about exactly what you are trying to achieve and then to look for a similar dilemma in the natural world. For example, if you want to find a way to cool a building down, look for large, complex structures in the hottest parts of the planet for inspiration – such as termite mounds in Africa. If you want to develop a paint that cleans itself, imagine another surface that remains free from dirt – the lotus plant, for example, whose leaves remain clean even in the midst of muddy swamps.

This approach to biomimicry has also delivered results for inventor Dean Cameron, the man behind the Biolytix household waste-water system. Cameron had two problems with the plastic tanks he used for the Biolytix systems. The first was that transporting completed tanks was difficult and expensive because they were so bulky, yet most of that bulk was air. Secondly, the tanks were built using rotational moulding, which is expensive. It would be cheaper to build the tank in parts, using injection moulding, but then he faced the challenge of tightly locking the parts together to form a complete tank.

What Cameron needed was a flexible and waterproof way of joining segments of tank together, so the tank could be produced in parts and then easily transported and assembled on-site. 'The plastic consultants that we got in to advise on how we might go about joining these large plastic parts, they said basically it can't be done economically – there was no way of doing it,' says Cameron. Others might have given up but, as Cameron says, 'that was like a red rag to a bull for me'. He figured that if it couldn't be done, then someone needed to invent a way.

Inspiration came from an unexpected source – a children's book. While reading to his son, Cameron learned about how the clam attaches itself securely onto rocks and crevices using only a fragile thin thread armed with tiny hooks. 'The principle is that the thread, even though it's very easy to bend like fishing line, when you pull it, it's actually quite strong,' he says. 'If nature uses it in mission-critical applications like that for the clam, then it will probably be pretty good for my application as well.'

This inspired Cameron with the solution for his problem, and that was Joinlox. Described as a sort of industrial velcro, Joinlox consists of two rows of intermeshing castellated hooks that can be locked together by means of a key – a flat, similarly castellated bar that inserts between the two rows of hooks and slides across to wedge the two parts firmly together. There are no bolts to tighten or screws to insert, and the whole assembly is flexible enough to be made from semi-rigid materials. Even better, it is strong and waterproof, perfect for water tanks.

Cameron has a track record of looking to nature for his solutions. In developing his Biolytix waste-water system, Cameron looked for an ecosystem in nature where large amounts of faecal matter were processed. He came across a bat colony. 'There are many thousands of bats in a concentrated roosting area and yet less than 300 metres downstream the water is crystal clear,' he says. The ecosystem achieves this impressive feat not just via the industrious actions of organisms such as worms, beetles and bacteria, but through the actions of trees and soil. The palm trees that flourish in this area are nutrient-loving species that suck up huge amounts of the nutrient load, while the soil itself provides another level of filtration.

Biomimicry doesn't always have to be a solo effort. Enterprises such as A skNature.org make the task of finding a solution in nature a whole lot easier. This open-source project, which Janine Benyus describes as the non-profit arm of the Biomimicry Guild, is intended as a wellspring of 'bioinspiration'. AskNatu re.org begins with the question 'How would Nature...', and it encourages users to contribute research insights, photos, illustrations or sketches that might provide an answer.

The site is organised into 'biological strategy' pages, each of which focuses on a particular solution from nature. The strategies are organised into what the site calls a 'biomimicry taxonomy'. The top level of insights include 'maintain physical integrity', 'get, store or distribute resources' and 'process information', which then break down further into more specific solutions. As kNature.org features more than 1500 biological strategy pages, covering a dazzling array of innovations. For example, there is a page dedicated to the mechanism by which the tubercules and bumps on the flippers of a humpback whale increased its aerodynamic efficiency; another focuses on how leaves being damaged by pests send out chemical signals that attract predators that eat the pests.[12]

Biomimicry is also becoming big business. The Biomimicry Guild, which Benyus co-founded, has established itself as a consultancy that helps companies and communities 'find, vet, understand and emulate life's time-tested strategies'. This means first identifying exactly what problem a company is trying to solve, then taking an 'amoeba through zebra' look at the natural world to see what solutions it has developed for that particular problem. The results tend to astound many of the guild's clients, who are often quite amazed by the innovative ways in which nature solves problems of ventilation or water storage, for example.

Inventor Kevin Inkster has become something of an accidental biomimic. Hailing from the suburb of Darlington in Perth, Australia, he declares himself to have been (and to still be, at heart) a hippie dropout. Despite this, he has found time to found a highly successful innovation company, Arbortech, which has produced over sixty cutting tools, as well as the Airboard personal hovercraft, which was used in the Opening Ceremony of the Sydney Olympic Games.

But look at Inkster's products and you see something interesting about them – many seem more than a little organic. Some of his saws look like they may have once been part of a giant insect, and other tools seem to have been modelled from bones. Surprisingly, this came about by accident rather than design. The company uses a fairly standard design software package to create its tools, but one of its features is that the user can specify that the design should be as minimalist as possible. 'The basic brief that I give the engineers is "I want you to take away every single thing that isn't necessary",' Inkster explains. 'We always start from that foundation.' His designers found that, with this brief, the design software began to produce some rather unusual things. 'You end up with things which are shaped more like animal than a machine,' Inkster says. 'Things that would ordinarily be struts like tubular steel will end up bone-shaped, narrow in the centre and wide near the edges.'

There is a simple explanation for this: when designing products, Inkster tries to use as few resources as possible. New modelling tools allow all of the desired product qualities to be put in – strength, torsion, compression – and then automatically create designs that meet these properties without wasting material. Designs like this are only copying the process that nature has been using for millions of years. As we continue to focus on resource efficiency, we will start to see even more similarities between what is natural and what is made by humans.

The results came as a surprise to Inkster, who came from a background of furniture building. 'We used to cut wood straight because straight is easy to do and easy to join, but with computers you can have any shape and exactly match that shape on another piece – you can have joinery that doesn't have to be straight,' he explains. 'Once you are removed from those constraints of having to do straight lines, curves and various geometric shapes – any shape is possible – then efficiency will dictate it will end up looking like a plant or animal form.' The Arbortech design process now almost intuitively starts with an animal shape before it even gets to the computer modelling phase.

With such a diversity of solutions, it's no wonder that so many innovators are taking a leaf out of nature's engineering textbook. But biomimicry is not limited to the design of objects. It is also providing inspiration on a much grander scale.

Nature the architect

The Indian region of Lavasa, nestled into the foothills of mountains 100 kilometres south-east of Mumbai, suffers some extremes of weather. It gets solidly drenched for three months during the monsoon season, but remains arid for the rest of the year. This poses a particular challenge for urban design, not to mention for management of resources such as water.

The global architecture firm HOK was given the challenge of designing an 8000-acre city in Lavasa. Faced with the issue of water management, the team from HOK turned to the region's original ecosystem – which, before the advent of intensive agriculture in the area, was most likely a forest – for inspiration. In partnership with engineering consultancy Buro Happold, they are designing buildings inspired by the area's original banyan trees. These trees are able not only to store water in their foundations but also to collect it via rooftops. The group is also taking lessons from ants on how to divert and channel water overflow away from areas where it is not wanted.[13]

Nature has provided ample inspiration for architecture over the course of human history. Ancient Egyptian columns were inspired by the shape and structure of the lotus flower and papyrus reed, and now modern designers are coming to appreciate the elegance, simplicity and functionality of nature's architecture.

The termite is one creature most architects and builders regard with loathing, due to its fondness for devouring wooden structures. One architect has a different view. When designing the Eastgate shopping centre and office building in Harare, Zimbabwe, Michael Pearce thought about the termites whose mounds dot the Zimbabwean savannah. These termite mounds are extraordinary in that they maintain a constant temperature inside the mound, even while the outside air temperature fluctuates enormously. They achieve this by means of vents, which enable cooler air to be drawn into subterranean chambers at the bottom of the mound while hot air is vented out the top. The mounds are also shaped and oriented to minimise their exposure to the sun during the hottest part of the day.

By copying this process, the Eastgate building uses similar passive-cooling techniques. Firstly, the exterior of the building has a large thermal mass, so it is able to absorb a considerable amount of heat without transmitting it to the interior offices. The outside is shaded by deep overhangs that keep as much sun as possible off the windows and walls. Secondly, as the outside temperatures cool off during the evening, warm air that has built up during the day is vented into the cooler external environment via a number of brick funnels. Cool air is drawn into the building and fanned through cavities in the flooring. These design elements mean that the Eastgate building uses just ten per cent of the energy consumed by conventional climate-controlled buildings.

The more we look to nature, the more it will be reflected in our urban design. Our cities will start to look more like the natural environment; buildings will be designed to track the movement of the sun, as flowers do, to collect the maximum solar energy or to control the temperature of the internal environment. Some will be covered with rooftop gardens and living roofs that will filter rainwater, keep the building cool and even provide a habitat for animals, while others will be covered with solar panels.

While nature might be a wonderful architect and designer, looking to nature doesn't only mean looking for solutions in nature and redesigning them for our purposes. Sometimes we can take advantage of these innovations in their natural habitat.

Partnering with nature

First there was peak oil. Now scientists are starting to talk about 'peak water' – the point at which our demand for fresh water exceeds its supply – or, more specifically, when it exceeds the rate at which the supply is replenished. Water covers more than seventy per cent of the Earth's surface, but the problem is that the vast bulk of this water is not fit for consumption, being too salty, dirty, contaminated or inaccessible. The solution to peak water is, therefore, to find a way of making this undrinkable water drinkable.

Water treatment is not a cheap process, as we saw when the city of New York began pricing the ecosystem service of the water filtration provided by the nearby Catskill Mountains. Current technologies force dirty water through membranes at high pressure, requiring significant amounts of energy, as well as considerable maintenance. It's costly not just financially but also in resource consumption and waste.

Water-filtration technology was a great leap forward, but in some cases we're just reinventing the wheel. Nature has been purifying water with underground aquifers for millennia. These are layers of water-permeable rock, or loose materials such as gravel and sand, from which groundwater can be extracted, according to Dr Simon Toze, principal research scientist with the CSIRO's Water for a Healthy Country Flagship. 'In layman's terms, an aquifer is an area below ground that is porous enough to hold water to the level that if we put a tube into the ground we can pull the water out – it holds enough that we can recover water from it,' he says.

At their simplest, aquifers are great places to set water aside for a non-rainy day. For example, water demand in Perth and Adelaide fluctuates significantly between the seasons, with low demand in winter and high demand in summer. Aquifers provide a much-needed storage facility for excess water during winter, which can then be used during the peak of summer. Water evaporation is also dramatically reduced below the ground – far more efficient than an above-ground dam.

Aquifers perform another service that is even more valuable: as well as storing water, they clean it. Aquifers are like naturally occurring underground water-treatment plants. The first form of filtration is physical – as the water moves through the aquifer, and particularly through sand aquifers, larger particles in the waste water are trapped. Aquifers also perform a chemical filtration, whereby aquifer materials such as limestone chemically interact with phosphates and other unwanted substances and bind to them. The third level of filtration is biological. 'There are native micro-organisms in aquifers, and a lot of these micro-organisms tend to be very adept at using a wide range of things for energy and growth,' says Toze. 'Pathogens and viruses are just bits of food.' The process takes a couple of months at least, but the end result is water that can be reused for irrigation and even, with further treatment, for drinking.

Managed Aquifer Recharge (MAR) is already being used in Perth to store and process treated sewage that would otherwise be discharged out to sea. The treated effluent is piped into a concrete chamber that sits above a row of horizontal trenches, or galleries, containing slotted pipes; these allow the waste water to filter down through the soil and into the aquifer. The water is finally recovered from the aquifer around seventy-five days after it is first injected into the site. This particular MAR project filters around fifty kilolitres per day – the volume of a typical home swimming pool – with the filtered water used to water green open spaces such as parks, golf courses and ovals.

Another CSIRO group is looking at capturing stormwater runoff and processing it through aquifers to produce water of drinking quality. While this process takes a lot longer – twelve months – and is filtered through a reed-bed filtration system first, the team has succeeded in getting potable water at the end of it, although it is slightly brackish. Not surprisingly, aquifer storage and recovery is also very economical. One report found that the average cost of water processed through eight urban stormwater aquifer projects was $1.12 per kilolitre, which is less than the current price of mains water in capital cities.

Another field benefitting from bio-partnerships is that of human-waste processing, where the humble earthworm is proving itself a very industrious partner. The earthworm's environmental virtues are hardly a new thing – their role in improving and maintaining soil fertility has been appreciated by gardeners around the world for millennia. But now they are being employed on a large scale, not only to process organic municipal waste but also the sludge left behind from the processing of sewage. Normally, these biosolids have to undergo extensive processing before their organic material can be reused – processing that uses valuable energy. However, Australian company Vermitech has developed large-scale worm farms that can process biosolids, converting them into high-value organic fertiliser with a minimum of dust, odour, noise and cost.

As we combine high technology with natural processes, we are finding something quite astounding: nature often beats current technologies in the efficiency stakes. It shouldn't come as a surprise. After all, nature has been perfecting its design skills over billions of years. We've only just started.

Natural systems and industrial ecology

If we take the study of nature a step further, we can learn something about the design of the whole system. Nature is a closed loop – nothing is wasted – so the organisation of natural systems is an ideal field we can look to for resource-efficiency opportunities. Understanding natural systems has given rise to a whole new discipline, that of *industrial ecology.* It was first mooted by Robert Frosch and Nicholas E. Gallopoulos, who, in a 1989 *Scientific American* article, posed several interesting questions: 'Why would not our industrial system behave like an ecosystem, where the wastes of a species may be resource to another species? Why would not the outputs of an industry be the inputs of another, thus reducing use of raw materials, pollution, and saving on waste treatment?'[14]

From an industrial ecologist's perspective, cities are just another large organism. And, as with an organism, you can identify different parts, or components, which interact with each other. Instead of the head, arms and torso, there are the commercial, industrial and residential districts. Instead of nerves and blood vessels, there are transportation mechanisms such as roads or powerlines. Waste streams are, well, waste streams. Changes in one part of the system can affect other parts, whether they are traffic jams or fluctuations in house prices. It's no coincidence that people talk about the 'beating heart of a city'. Only when you start to treat the city as having its own ecology can you start to comprehend and manage its complexity.

Industrial ecology is particularly useful when you're designing production and consumption processes that aim to minimise the waste produced and the environmental impact. It is driven by engineering innovations that increase material and energy efficiencies of specific processes, and by a need to take a whole-of-system perspective, which identifies ways for unwanted by-products and waste to be avoided or used elsewhere in the system. Industrial ecology brings together ideas such as life-cycle assessments, material and energy flow studies, product stewardship and eco-design.

In this way, industrial ecology can give us insights into the similarities between processes in the biosphere and those in the technosphere. The technosphere exists within something called the *market* – just as the biosphere exists in the broader *environment.* From biosphere to technosphere, *organisms* become *companies, natural selection* becomes *competition,* and *adaptation* becomes *innovation.* Even concepts such as a *food web* translate to the technosphere as a *product lifecycle.* When we look at things this way, we can learn how to create more efficient systems.

Taking an ecological approach can also give us insights into the way we create businesses and societies. Businesses are now seeing the value of creating internal 'ecosystems' around their products and services, and are shifting away from a top-down, proprietary way of doing things and towards a more organic approach. The same is happening on the internet, with what is known as *Web 2.0.* In the past, online business directories were always structured as taxonomies, with a structure imposed on the directory which businesses had to fit. Now we are seeing the rise of the *folksonomy* in directories like HotFrog, where businesses categorise and tag themselves, using their own words. This builds up a far more comprehensive and accurate taxonomy, which is able to grow and adapt to the 'organisms' within it. This sort of approach requires little intermediation: people collaborate directly within and across organisational boundaries.

The natural world is complex, and this complexity makes it highly resilient. For example, if a forest were to rely on only one plant species to perform the vital function of fixing nitrogen into the soil, it would be vulnerable to any event that might wipe out that one species, such as the introduction of a particular virus. However, if there were ten nitrogenfixing species, at least a few are likely to survive a catastrophe, ensuring that the forest has a much better chance of recovery. Similarly, mass collaboration in human endeavours leads to more reliable, resilient outcomes.

Wikipedia is a classic example of mass collaboration, and the approach it uses has proven itself to be effective. A recent study found similar levels of accuracy between Wikipedia and the *Encyclopaedia Britannica.* As James Surowiecki outlines in his wonderful book *The Wisdom of Crowds,* what Wikipedia did was to create a whole ecosystem around its information, creating the right incentives for group participation. Rules were created to solve disputes and key contributors were rewarded through recognition of their contributions. Wikipedia even evolved much like a biological ecosystem.[15]

Businesses are taking lessons from the operation of complex societies in nature. Swarm intelligence describes the collective behaviour of a group of individual agents – it governs the behaviour of organisms such as ants, bees and wasps. According to Eric Bonabeau and Christopher Meyer in the *Harvard Business Review,* [16] one of the reasons these collective societies are so successful across the planet is not only because they are flexible and resilient, but specifically because their activities are self-organised – they are neither controlled by a central agency nor locally supervised.

By taking a 'swarm intelligence' approach to problems, companies are coming up with some innovative and efficient solutions. For example, Bonabeau and Meyer note how Hewlett-Packard developed software to manage congestion on telecommunications networks that takes inspiration from ants' foraging techniques. Ants direct other ants to food sources by leaving pheromone trails, so Hewlett-Packard developed software that sent digital 'agents' roaming throughout a telecommunications network looking for less congested pathways and leaving the digital equivalent of a pheromone trail to mark those paths. Phone calls could then follow those trails. The program even specifies that the trails 'evaporate' after a certain period of time so the program can adjust to changes in traffic.

Societies have also evolved to consider how to ensure the long-term viability of a community at the same time as that of the individuals within that society. Tax codes try to strike a balance between encouraging entrepreneurialism while also creating benefit for society as a whole. Legal systems penalise those who use resources at the expense of the environment and society. Perhaps future policy-making will also learn from natural systems?

An organic world

There is one final reason that smart businesses are looking to nature: their customers are doing it. This is understandable when you think about the components of many of the products we buy and the foods we consume; many household products contain toxic metals and chemicals that are used in the production process. Pesticides can leave trace materials in the foods that we eat. Building products can contain dangerous substances that have long-lasting health effects, as our use of asbestos has tragically demonstrated.

Endocrine-disrupting chemicals are a scourge of the modern world. They mimic the hormones that control development and function in our bodies, and they have the potential to disrupt hormonal systems in many other species too. While we are now well aware of the potential harm of endocrine-disrupting chemicals such as Bisphenol A and phthalates, they are still found in everything from vinyl flooring and soap to inflatable toys and computers.

Consumers are also becoming increasingly aware of the impact of toxic substances on the environment. Excessive use of fertilisers is causing major environmental problems, particularly in our rivers, lakes and marine areas. The Great Barrier Reef has experienced a thirty per cent jump in the amount of phosphorus and nitrogen coming to it from the mainland, as nearby catchment areas are developed for agricultural uses.[17] This causes excessive phytoplankton growth, which cuts down the light available to coral and seagrass communities. Tubeworms, sponges and barnacles love these increased nutrients; their numbers are skyrocketing and they are out-competing coral. Algae also flourish in the nutrient-rich environment, growing over the coral that remains and effectively suffocating it. Finally, phosphorus can weaken coral skeletons, and herbicides can affect the health of the seagrass that feeds dugongs. It's a bleak and distressing scenario for one of the natural wonders of the world.

Consumers' environmental consciences are governing their wallets more and more, and we're seeing increasing demand for more environmentally responsible products in our supermarkets. UK supermarket Sainsbury's organic range is growing by 18.4 per cent year on year.[18] The global market for organic food and drink exceeded US$40 billion in 2006 – one-third of which was as fresh produce – and it continues to grow rapidly. The US natural and organic food retailer Whole Foods Market began life in 1980 with just one small store in Austin, Texas, employing nineteen people. The chain has now grown to become the world's largest retailer of organic food, which it sells through 270 outlets across North America and the United Kingdom.

The fundamental principle underlying organic food production is that it does not use any form of synthetic fertiliser, pesticide, growth enhancer or feed additive, nor does it allow any genetic modification. These products and processes are all relatively new arrivals on the agricultural scene, meaning that much of organic farming is about going back to basics and learning from natural systems. For example, instead of artificial fertilisers, organic farmers use compost, green manure and crop rotation to ensure that nutrients in the soil are carefully managed and replenished. Instead of toxic pesticides, organic farming uses techniques such as companion planting – planting species that are naturally pestresistant or repellent – and by introducing and encouraging natural predators that prey on pests.

Organic livestock is raised without the use of hormones or antibiotics, which, in some cases, has meant that farmers are turning back to 'heirloom' breeds that were once bred for specific desirable traits, such as lush wool, high milk production or beautifully marbled meat. Foods claiming to be organic must have no food additives and must not have been subject to processes such as chemical ripening or food irradiation.

While there is no doubt that the widespread use of pesticides and fertiliser helped farmers to feed a high proportion of the world's population, we are now looking at how we might get the same benefits without the downside. When you add this aim to the efforts to reduce waste and carbon emissions, it's clear that in the sixth wave we will see a world that is a lot more natural.

A natural future

We tend to think that we as a species somehow stand apart from nature. Indeed, we often see ourselves as struggling against nature for our survival. However, as technology develops and our population continues to grow, we are gaining a new awareness that our survival is entirely dependent on the natural world. Even more sobering is the realisation that, when compared to nature, we are novices in the presence of a profound genius.

Janine Benyus believes we are coming through what she thinks of as our 'toddlers with matches' period, where we were a little too enchanted by our own prowess. 'We still are, but increasingly we're realising that we've painted ourselves into a corner,' she says. The breakthroughs of this toddler period are *maladaptive,* in the sense that we've been poisoning our world. In the biological world, this propensity would very quickly be edited out of the gene pool but – fortunately or unfortunately – we are in a long feedback loop that may yet give us time to change our ways.

Changing our perspective can have a profound effect, particularly when we go back to our roots and realise that we are all part of a natural system. Like errant children, we are coming to realise that we can learn something from our parents. And, like all caring parents, Mother Nature has many lessons for us. First and foremost, we must create less waste – nature is a closed loop and doesn't use any more resources than she absolutely has to. Nature is also ingeniously elegant and has many tricks up her sleeve, some of which we are only now starting to learn. She is also forgiving and wants to work with us to fix our problems, if only we will let her.

Epilogue

Eco-natives

In 2001 the American author Marc Prensky coined the term *digital natives* to describe the generation of youngsters who have grown up never knowing a world without computers or the internet. This generation instinctively understood the lessons of the fifth wave, from the need to constantly embrace technology to the power of networking and information.

The sixth wave will see a new generation of children, who will understand the basic lessons of this book in a fundamental way. These are the kids who will grow up with environmental programs in their schools, who will admonish their parents to turn off the tap while brushing their teeth, for whom recycling and composting are second nature, and whose Hollywood role models drive electric vehicles and champion environmental causes. They will be the *eco-natives.*

At the start of this book we met Allister and his son, William. Like all babies growing up in this and the next decade, William will have a relationship with the world that is completely different to his father's. He will grow up in a world of limits – limits on energy, limits on resources and limits on waste. But this world will also be one of extraordinary opportunity and innovation.

It will be a world where everything has a value and nothing is discarded or wasted, where the air and water are cleaner, where the destruction of wilderness is slowly turned around and where the natural environment becomes a thing to be cherished and nurtured.

It will be a world where the boundaries that might once have restricted opportunities are dissolved, so that people are no longer constrained by geographic distance. Instead, they will have become one interconnected organism. And in this new world, the harder edges of human civilisation will soften; people will become more organic, as they finally cease their fruitless battle against nature and start to embrace her wisdom, guidance and inspiration.

This book has been about industrial innovation and the major changes that we might see in the future. At the heart of the sixth wave of innovation is a fundamental shift – for the first time since the Industrial Revolution, the first wave, we are about to decouple economic growth from resource consumption. It is a fire that has been smouldering for some time but will finally burst into life, sparked by resource shortages and pollutants, fuelled by institutional changes like carbon pricing, and accelerated by clean technologies.

Eco-natives will understand this new world. They will look for waste as a source of opportunity. They will seek out services rather than products, and will see little distinction between the digital and natural worlds. Eco-natives will use services globally but will prefer local products, and they'll look to nature for inspiration. They will wonder how we ever lived any other way, in the same way that digital natives find it hard to imagine a world without email.

You may be thinking that this paints a very positive view of the world. It does, but of course it all depends on the correctness of our prediction that the next wave of innovation – the sixth wave – will be based on resource efficiency. We could be wrong.

At the beginning of the Industrial Revolution, a prediction was made by a fellow called Thomas Robert Malthus. He predicted that the rate of technological progress would always be outstripped by population growth. According to Malthus, humanity would always be limited by its access to resources, and technological growth would always be matched by population growth, so that any gains made through technological improvements would simply go towards conceiving and feeding more people.

In Malthus' view, humanity would never escape this cycle. 'The power of population is indefinitely greater than the power in the earth to produce subsistence for man,' he wrote.[1] Humanity would remain at a subsistence level, just as it had done for the centuries before. Famine and disease would be forever nipping at our heels, no matter how hard we strived to outwit it. Unfortunately for Malthus – and fortunately for us – his prediction did not eventuate. Humanity did escape the population trap and has significantly increased living standards for many (but not yet all) of its citizens. Pessimists might say that this is only temporary. In a world that will host nine billion people in 2050, perhaps we are just enjoying a brief reprieve from Malthus' dire prediction?

There is a solution, though. The Harvard economist Michael Kremer believes that the rate of change of technological growth is exponential and is proportional to the number of brains on the planet – in other words, the more people we have, the faster we innovate. For this reason, Kremer believes, we've been able to come up with technologies and inventions to improve our standard of living faster than we have seen the population rise.[2]

This all leads to a critical challenge. As the equation I=PAT states, our impact on the environment is a function of our population, our affluence and the technologies available to us. So, while populations grow and affluence increases, we run the risk of severely impacting our natural resources. One of the hallmarks of waves of innovation, however, is that technology should never be underestimated; if the next wave of innovation is centred on resource efficiency, perhaps the technology will win out without compromising either population or affluence.

The social ecologist Peter Drucker once said, 'The best way to predict the future is to create it.' In the spirit of that sentiment, we are setting out to create a new future. We believe that the sixth-wave trends we have outlined in this book are compelling, but not nearly as compelling as the environmental issues that will plague us should we not act quickly enough.

The sixth wave is a call to action, and we encourage all of the world's innovators, entrepreneurs, business people and individual consumers to take part. We do this not for ourselves, but for our children and our children's children – the Williams, Jaspers and Ninas of this world – who must live, survive and thrive in the world we leave them.

Final musings

The seventh wave

You might be wondering – if the sixth wave is going to be based on resource efficiency, in which direction could the seventh wave of innovation take us?

If our prediction is correct, by the end of the sixth wave we will have an economy in which economic growth is no longer dependent on the irreversible consumption of resources. The creation of value will be linked to human time and ingenuity, and intellectual property will take on a new significance. In this scenario, innovation might well be focused on extracting the most out of human resources: the seventh wave could be the human efficiency or human capacity wave.

Humankind would join with computers through implants and cybernetic enhancements. The investments of the ageing baby-boomers in medical research would have left us many lifestyle advances that ensure we keep working longer, better and smarter. We would even move beyond the human genome project into mapping the human brain – the last frontier of complexity. The key drivers of technology will be getting people as smart and as healthy as possible.

With people living longer and longer, where could future growth come from? Now we're really speculating, but let's have some fun. Perhaps growth in the eighth wave will come from the mass colonisation of outer space.

For more thoughts on the matter, or to have your say, check out our website: www.sixthwave.org.

Acknowledgements

A book looking to the future doesn't come together without support and inspiration from many sources.

Thank you to Meredith Curnow, Elizabeth Cowell and Margaret Seale at Random House Australia for their patience and encouragement and for believing in this book, and to editor Julian Welch for his diligence, good advice and ego-boosting feedback.

Many thanks also to our reviewers and mentors, who have all helped to shape the contents of this book and made it even better than the authors could have possibly achieved alone. In particular, we would like to thank Geoff Garrett, Mark Dodgson, Greg Bourne, Tony D'Aloisio and Terry Cutler for their support and guidance over many years.

We would also like to acknowledge and pay tribute to all of the people who we talked to in writing this book: the entrepreneurs, thought leaders and innovators at the vanguard of the next wave. Their passion and commitment to creating the future are something amazing to behold.

And finally, on a personal note...

James: This book would not have occurred without the prompting of many people who saw my original talk on the next wave of innovation and asked me to expand on it – thank you all for being so persistent. Thank you to my wonderful wife, Geraldine, for supporting, challenging, inspiring and loving me – you give me the courage to be the best I can be – and also to my new son, Jasper, for having the final draft of the book read to him when he was just weeks old. Big thanks also to my parents, Lynne and Miles, who have always encouraged me to look to the future, to my tireless co-author, and to my friends and family, from whom I learn something new every day.

Bianca: Some might question the wisdom of attempting to write a book shortly after giving birth to my first child – the gorgeous Nina – but thankfully I never asked the question and simply sailed into uncharted waters secure in the support of my family, friends and co-author. I owe a huge thank you and lots of cupcakes to my mothers' group in the Blue Mountains who have all been so wonderfully encouraging, supportive and saved my sanity on so many occasions. Thank you to my parents, Kerrie and Stephen, my brother, Matty, and sister, Lucy, for their love and support and for being fabulous. Thank you to my little Nina for being a baby who does sleep well during the day, even if she doesn't at night. And finally, thank you to my beloved husband, Phil, for giving me the time and space to be able to write this, and for believing in me, supporting me, looking after me and loving me.

About the authors

James Bradfield Moody is Executive Director, Development, at Australia's national research agency, CSIRO. James also sits on the Advisory Council of the Australian Bureau of Statistics, is a board member of the Brisbane Institute in Queensland, an Australian national commissioner for UNESCO and a World Economic Forum Young Global Leader. He has also worked with the United Nations Environment Programme, was Executive Secretary of the UN Taskforce on Science and Technology for the Millennium Development Goals and has held board positions with the Australian Government Bureau of Meteorology and the National Australia Day Council. James received his PhD in innovation theory from the Australian National University and was a chief systems engineer for the Australian Satellite FedSat. He is a regular panellist on the ABC television show *The New Inventors* and lives in Sydney, Australia, with his wife and son.

Bianca Nogrady is a freelance science journalist and broadcaster who has written regularly for publications such as *Scientific American, The Australian* newspaper, *G Magazine* and the Australian Broadcasting Corporation science and health websites. For several years she was the radio voice of *New Scientist* magazine in Australia and was also a medical journalist with *Australian Doctor* newspaper and founding editor of the CSIRO children's science magazine *Scientriffi c.* Bianca lives in the Blue Mountains, Australia, with her husband and daughter.

Notes

Chapter 1: Why do things keep changing?

[1]   Life expectancy trends can be found at: www.aihw.gov.au/mortality/life_expectancy/trends.cfm.

[2]   For a good background on different types of innovation, see: Dodgson, M., *The Management of Technological Innovation: An International and Strategic Approach*, Oxford University Press, Oxford, 2000.

[3]   Freeman, C., 'The Economics of Technical Change', *Cambridge Journal of Economics*, vol.18, 1994, pp463–514.

[4]   North, D., 'Institutions', *The Journal of Economic Perspectives*, vol.5 (winter 1), 1991, pp97–112.

[5]   For a full discussion of the drivers behind and impacts of long waves, see Freeman, C., Louca, F., *As Time Goes By: From the Industrial Revolutions*

to the Information Revolution, Oxford University Press, Oxford, 2001; and Perez, C., *Technological Revolutions and Financial Capital: The Dynamics of Bubbles and Golden Ages,* Edward Elgar, Cheltenham, UK, 2002.

[6] Papenhausen, C., 'Causal Mechanisms of Long Waves', *Futures,* vol.40, 2008, pp788–794.

[7] Freeman, C., Louca, F., *As Time Goes By: From the Industrial Revolutions to the Information Revolution,* Oxford University Press, Oxford, 2001.

[8] This efficiency was achieved through catalytic cracking, invented by William Meriam Burton in 1912. For a history of William Burton, see 'Inventor of the Week: William Burton', Lemelson MIT Program, Massachussets Institute of Technology, August 2008: http://web.mit.edu/invent/iow/burton.html.

[9]   Chris Papenhausen is a innovation theorist at the University of Massachusetts, Dartmouth. 10. Sterman, J., *An Integrated Theory of the Economic Long Wave,* Massachusetts Institute of Technology, Cambridge, Massachusetts, 1984.

[11]   See, for example, Perez, C., *Technological Revolutions and Financial Capital: The Dynamics of Bubbles and Golden Ages,* Edward Elgar, Cheltenham, UK, 2002.

Chapter 2: The last wave

[1]   Intel's Chief Technology Officer, Pat Gelsinger, believes that Moore's law will continue at least ten to fifteen years following recent breakthroughs and sees it continuing to at least 2029. Geelan, Jeremy, 'Moore's Law: "We See No End in Sight," Says Intel's Pat Gelsinger', *Java Developer's Journal,* 1 May 2008:

http://java.sys-con.com/node/55
7154.

[2]   A brief history of the Intel 4004
processor can be found at: www
.intel.com/museum/archives/400
4facts.htm.

[3]   IBM gives a brief history of the
development of the Mark I at:
www-03.ibm.com/ibm/history/his
tory/decade_1940.html.

Chapter 3: Resource efficiency: the next great market

[1]   The US Department of Energy
website gives a full breakdown
of fuel economy for a range of
different types of vehicles: www
.fueleconomy.gov/FEG/atv.shtml.

[2]   Abstract established by Eduard
Pestel. Meadows, D.H., Meadows,
D.L., Randers, J., Behrens, W.W.,
*The Limits to Growth: A Report
for the Club of Rome's Project
on the Predicament of Mankind,*
Universe Books, New York, 1972.
3. Ibid, p.1. 4. Pimm, S.L., *The*

World According to Pimm: A Scientist Audits the Earth, McGraw-Hill, New York, 2001.

[5]   Cotton is only one of the products evaluated by the Water Footprint Network: www.waterfo otprint.org/?page=files/productga llery.

[6]   For a full lifecycle analysis of the computer chip, see 'Computer Chip Life Cycle', The Environmental Literacy Council: www.enviroliteracy.org/article.ph p/1275.html.

[7]   Hawken, P., Lovins, A. and Lovins H., *Natural Capitalism: The Next Industrial Revolution,* Earthscan, London, 1999. Amory Lovins gave a summary of the ideas presented in this book in a lecture entitled 'Natural Capitalism', given in Sydney on 4 July 2000. An edited version of his lecture is available at: w ww.abc.net.au/science/slab/natca p/default.htm.

[8]   For a history of platinum, see 'Platinum – History and Investment' at the A to Z of

Materials website: www.azom.com/details.asp ?ArticleID=1210#. See also: www.mineralszone.com/minerals/platinum.html.

[9]   A range of interesting food waste statistics can be found at: www.foodwise.com.au/Did%20you%20Know/Food%20Facts.aspx.

[10]   'A Hill of Beans', *The Economist,* 26 November 2009.

[11]   Waste and Recycling Action Program (WRAP): www.wrap.org.uk/retail/food_waste/nonhousehold_food.html.

[12]   Full details of CAPTCHA and reCAPTCHA can be found at: www.captcha.net. You can even generate your own CAPTCHAs using tools at this site.

[13]   Von Ahn, L., Maurer, B., McMillen, C., Abraham, D. and Blum, M., 'reCAPTCHA: Human-Based Character Recognition via Web Security Measures', *Science,* vol.321, no.5895, 12 September 2008, pp1465–1468.

Chapter 4: Capturing true costs: evolving institutions

[1]   Food and Agricultural Organisation, Global Forest Resource Assessment 2005: ftp: //ftp.fao.org/docrep/fao/008/A04 00E/A0400E01.pdf.

[2]   Nellemann, C., Miles, L., Kaltenborn, B.P., Virtue, M. and Ahlenius, H. (eds), 'The Last Stand of the Orangutan – State of Emergency: Illegal Logging, Fire and Palm Oil in Indonesia's National Parks', United Nations Environment Programme, GRID-Arendal, Norway, 2007: w ww.grida.no/publications/rr/oran gutan.

[3]   Ehrlich, P.R. and Holdren, J., 'Impact of Population Growth', *Science*, vol.171, no.3977, 26 March 1971, pp1212–17.

[4]   Data on a whole range of growth and population trends can be found at the World Resources Institute EarthTrends website: h ttp://earthtrends.wri.org.

[5]   The Blacksmith Institute reports on the world's most polluted places. Linfen was featured in the 2007 report: www.worstpoll uted.org/projects_ reports/display/22.

[6]   World Health Organization air quality guidelines can be found at: www.who.int/phe/health_topi cs/outdoorair_aqg/en/index.html.

[7]   Chang, E., 'Choking in China's Polluted City', CNN, 16 December 2009: http://edition.cnn.com/20 09/WORLD/asiapcf/12/15/china.p ollution.

[8]   'Report by the Director of Meteorology on the Bureau of Meteorology's Forecasting and Warning Performance for the Sydney Hailstorm of 14 April 1999', available at: www.bom.g ov.au/inside/services_policy/stor ms/sydney_hail/hail_report.shtml .

[9]   Schuster, S.S., Blong, R.J., Leigh, R.J. and McAneney, K.J., 'Characteristics of the 14 April 1999 Sydney Hailstorm Based on Ground Observations, Weather

Radar, Insurance Data and Emergency Calls', *Natural Hazards and Earth System Sciences,* vol.5, 11 August 2005, pp613–20.

[10]   Coleman, T., 'The Impact of Climate Change on Insurance against Catastrophes', Insurance Australia Group, available at: www.iag.com.au/sustainable/media/presentation-20021219.pdf.

[11]   The Center for Health and the Global Environment at Harvard Medical School has conducted a detailed analysis of the health, ecological and economic dimensions of climate change. The 'Climate Change Futures' report can be downloaded from: www.climatechangefutures.org/report. For a discussion of the financial implications of climate change for the insurance industry, see page 92.

[12]   The United States Environmental Protection Agency gives a detailed overview of the Acid Rain

Program at: www.epa.gov/NE/e co/acidrain/allowances.html.

[13] Stern, N., 'Climate Change, Ethics and the Economics of the Global Deal' (lecture), Royal Economic Society, November 2007: www.res.org.uk/society/l ecture2007.asp.

[14] The text of the Kyoto Protocol can be found at: http://unfccc. int/resource/docs/convkp/kpeng .pdf. Information on trading mechanisms can be found at: http://unfccc.int/kyoto_protocol /mechanisms/items/1673.php.

[15] 'State and Trends of the Carbon Market 2009', The World Bank, available at: http: //wbcarbonfinance.org/Router.c fm?Page=DocLib&Catalog ID=47687.

[16] 'Reducing Carbon Emissions from Deforestation in the Ulu Masen Ecosystem, Aceh, Indonesia: A Triple-Benefit Project Design Note for CCBA Audit', Provincial Government of Nanggroe Aceh Darussalam, 2007: www.climate-standards.o

rg/projects/files/Final_Ulu_Mase n_CCBA_project_design_note_D ec29.pdf.

[17]   *Ecosystems and Human Well-being: Synthesis,* Millennium Ecosystem Assessment, Island Press, Washington, DC, 2005.

[18]   More information about the Catskills Water Corporation can be found at: www.cwconline.or g/about/ab_index.html.

[19]   A case study of the New York City Watershed Partnership can be found at: www.epa.gov/nce i/collaboration/nyc.pdf.

[20]   Chichilnisky, G., and Heal, G., 'Economic Returns from the Biosphere – Commentary', *Nature,* vol.391, 12 February 1998, pp629–30.

[21]   Hardin, G., 'The Tragedy of the Commons', *Science,* vol.162, no.3859, 13 December 1968, pp1243–48.

[22]   Ibid., p.1244, reprinted with the permission of AAAS.

[23]   More details on the work of Hernando de Soto can be found

in Clift, J., 'Hearing the Dogs Bark', *Finance & Development,* vol.40, no.4, December 2003, available at: www.imf.org/external/pubs/ft/fa ndd/2003/12/pdf/people.pdf.

[24]   Costanza, R., et al., 'The Value of the World's Ecosystem Services and Natural Capital', *Nature,* vol.387, 1997, pp253–260.

[25]   Zebich-Knos, M., 'Preserving Biodiversity in Costa Rica: The Case of the Merck-INBio Agreement', *The Journal of Environment & Development,* vol.6, 1997, pp180–186.

[26]   More information on INBio can be found at: www.inbio.ac.cr/e n/inbio/inb_queinbio.htm.

[27]   Zebich-Knos, M., 'Preserving Biodiversity in Costa Rica: The Case of the Merck-INBio Agreement', *The Journal of Environment & Development,* vol.6, 1997, pp180–186.

[28] Globescan, Corporate Social Responsibility Monitor: www.globescan.com/csrm_overview.htm.

[29] Hardy, C., and Van Vugt, M., 'Nice Guys Finish First: The Competitive Altruism Hypothesis', *Personality and Social Psychology Bulletin,* vol.32, no.10, 2006, pp1402–1413.

[30] Goldstein, N., Cialdini, R., and Griskevicius, V., 'A Room with a Viewpoint: Using Social Norms to Motivate Environmental Conservation in Hotels', *Journal of Consumer Research,* vol.35, no.3, October 2008, pp472–482.

[31] Information about the Dow Jones Sustainability Indexes is available at: www.sustainability-index.com/07_htmle/indexes/overview.html.

[32] See the review of ten years of the Dow Jones Sustainability Index at: www.sustainability-index.com/djsi_pdf/news/PressRe

l e a s e s / S A M _
PressRelease_090910_DJSI_10yrs.pdf.

Chapter 5: Cleantech: from dotcom to wattcom

[1]   'The World's Billionaires: #396 Shi Zhengrong', *Forbes,* 5 March 2008: www.forbes.com/lists/200 8/10/billionaires08_Shi-Zhengron g_EP46.html.

[2]   In 2007 Zhengrong Shi was named one of *Time* magazine's 'Heroes of the Environment'. Green, M., 'Shi Zhengrong', *Time,* 17 October 2007: www.ti me.com/time/specials/2007/artic le/0,28804,1663317_1663322_1 669932,00.html.

[3]   The full report, 'Unlocking Energy Efficiency in the US Economy', can be found at: www.mckinsey .com/clientservice/electricpowern aturalgas/US_energy_efficiency.

[4]   US Department of Energy, 'Congestion and Transmission Losses on the US Power Grid': http://sites.energetics.com/gridw orks/grid.html.

[5]   The 2010 Annual Energy Outlook is produced by the US Energy Information Administration: www.eia.doe.gov/oiaf/aeo/demand.html.

[6]   International Energy Association: www.iea.org/textbase/papers/2005/standby_fact.pdf.

[7]   'Fleet Tracking Boosts Bottom Line for Service Today', *Asset Management News:* www.assetmgmtnews.com/content/view/307/6.

[8]   The wind data and economic data on Kite Power is taken from the Kite Gen Website at: www.kitegen.com/pages/winddata.html.

[9]   More information on the solar concentrator can be found at: solar.anu.edu.au/research/linearconc.php.

[10]   For information about the Desertec project, see: www.desertec.org.

[11]   Silicon Valley venture capital investment data is sourced from the Silicon Valley Index: www.jointventure.org.

[12]   For industry data and analysis, the Cleantech Group has been

tracking the cleantech industry since 1999: www.cleantech.com . 13. National Venture Capital Association, Venture Capitalists Predictions for 2010: www.nvc a.org.

[14] McKinsey & Company, 'How the World Should Invest in Energy Efficiency', *McKinsey Quarterley*, July 2008: www.mckinseyquart erly.com/Economic_Studies/Pro ductivity_Performance/How_the _world_should_ invest_in_ener gy_efficiency_2165.

[15] 'The World's Biggest Companies', *Forbes*, 8 April 2009: www.forbes.com/2009/0 4/08/worlds-largest-companies-business-global-09-global_land. html.

[16] An overview of GE's investments in clean technology is given at: http://ge.ecomagin ation.com/our-commitments/inv est-in-research.html. An interview with Jeff Immelt can be found at: http://news.bbc.c o.uk/2/hi/business/4204194.stm .

[17]   Google's plans in the renewable energy domain are outlined at: www.google.com/intl/en/press/pressrel/20071127_green.html.

[18]   The Australian Ethical Charter is given at: www.australianethical.com.au/ethical-charter.

[19]   A visual summary of the investment is found at: www.newscientist.com/data/images/archive/2698/26984001.jpg.

[20]   'How Energy Research is Going,' *The Economist,* 26 March 2009.

[21]   Advanced Research Projects Agency-Energy (ARPA-E): arpa-e.energy.gov.

[22]   'China's Clean Revolution II: Opportunities for a Low Carbon Future', The Climate Group, August 2009, available at: www.theclimategroup.org/publications/2009/8/20/chinas-clean-revolution-ii-opportunities-fora-low-carbon-future.

23.   The fund is detailed at Sustainable Development Technology Development Canada: www.sdtc.ca/en/funding/index.htm.

Chapter 6:
Waste=opportunity

[1]   Details of the ENERFISH project can be found at the European Union Seventh Framework Program website. Go to http://cordis.europa.eu/fp7/projects_en.html and then search for 'ENER FISH'.

[2]   Storm Brewing is a brewery in Newfoundland, Canada: www.stormbrewing.ca/STORM%20BREWING/One%20Little%20 Brewery.html.

[3]   A case study of reusing waste from the brewing process is given at: www.zeri.org/case_studies_beer.htm.

[4]   White, S., and Cordell, D., 'Peak Phosphorus: the Sequel to Peak Oil,' Sustainable Phosphorus Futures: http://phosphorusfutures.net/peakphosphorus.

[5]   For information on standby power and other green living tips, visit: www.greenlivingtips.c

om/articles/93/1/Standby-power-electricityconsumption.html.

[6]   Information about global transportation emissions can be found on the Planet Green Website: http://planetgreen.disc overy.com/tech-transport/plane-t rain-automobile-travel.html.

[7]   Hawken, P., Lovins, A., and Lovins H., *Natural Capitalism: The Next Industrial Revolution,* Earthscan, London, 1999, p 116.

[8]   Dr Karl-Henrik Robèrt's exploration of the link between cancer and sustainability can be read at: www.thenaturalstep.org /en/our-story.

[9]   Full details of the four Natural Step system conditions can be found at: www.naturalstep.org/t he-system-conditions.   In summary, they are that in a sustainable society, nature is not subject to systematically increasing concentrations of substances extracted from the earth's crust; or to concentrations of substances produced by society; or to

degradation by physical means; also, in that society, people are not subject to conditions that systemically undermine their capacity to meet their needs.

[10]   The twelve principles according to the Centre for Green Chemistry are:

1) It is better to prevent waste than to treat or clean up waste after it is formed.

2) Synthetic methods should be designed to maximise the incorporation of all materials used in the process into the final product.

3) Wherever practicable, synthetic methodologies should be designed to use and generate substances that possess little or no toxicity to human health and to the environment.

4) Chemical products should be designed to preserve efficacy of function while reducing toxicity.

5) The use of auxiliary substances (e.g., solvents, separation agents, and so forth) should be made unnecessary wherever possible and innocuous when used.

6) Energy requirements should be recognised for their environmental and economic impacts and should be minimised. Synthetic methods should be conducted at ambient temperature and pressure.

7) A raw material or feedstock should be renewable rather than depleting wherever technically and economically practicable.

8) Unnecessary derivisation (b l o c k i n g g r o u p, protection/deprotection, temporary modification of physical or chemical processes) should be avoided whenever possible.

9) Catalytic reagents (as selective as possible) are superior to stoichiometric reagents.

10) Chemical products should be designed so that at the end of their function they do not persist in the environment and break down into innocuous degradation products.

11) Analytical methodologies need to be developed further to allow for real-time in-process monitoring and control before the formation of hazardous substances.

12) Substances and the form of a substance used in a chemical process should be chosen so as to minimise the potential for chemical accidents, including releases, explosions and fires.

Sourced from the Centre for Green Chemistry: www.chem.mona sh.edu.au/green-chem/docs/cgc-rep ort-2008.pdf, p.26.

[11]   Dunn, P., Galvin, S., and Hettenbach, K., 'The Development of an Environmentally Benign Synthesis of Sildenafil Citrate (Viagra) and its Assessment by Green Chemistry Metrics', *Green Chemistry,* vol.6, 2004, pp43–48.

[12]   The full report, 'Unlocking Energy Efficiency in the US Economy', can be found at: w ww.mckinsey.com/clientservice/ electricpowernaturalgas/US_ene rgy_efficiency.

[13]   Ozaki, R., 'Adopting Sustainable Innovation: What Makes

Consumers Sign up to Green Electricity', *Business Strategy and the Environment,* Wiley Interscience, 3 July 2009.

[14]   For an overview of SETI@home, or to download the software to participate, visit: http://setiath ome.berkeley.edu.

Chapter 7: Sell the service, not the product

[1]   For more information on services, see the UK Royal Society study on the role of science in services sector innovation: http://royalso ciety.org/Hidden-wealth-The-cont ribution-of-science-to-service-sec tor-innovation.

[2]   Capehart, B. (ed.), *Encyclopedia of Energy Engineering and Technology,* CRC Press, Boca Raton, 2007.

[3]   Fox-Penner, P.S., *Return of the Energy Services Model: How Energy Efficiency, Climate Change, and Smart Grid Will Transform American Utilities,* The Brattle Group, Inc., May 2009.

[4]   For information about the uptake of technologies when the utility provider is involved in the process, see Fox-Penner, P.S., *Return of the Energy Services Model: How Energy Effi ciency, Climate Change, and Smart Grid Will Transform American Utilities,* The Brattle Group, Inc., May 2009.

[5]   European Union Framework Program 7 – Software & Service Architectures and Infrastructures: http://cordis.europa.eu/fp7/ict/ss ai/home_en.html.

[6]   Tax Norway receives personal information used to pre-complete tax returns from a number of sources, including employers and banks: www.skatteetaten.no/en/ taxnorway.

Chapter 8: Digital and natural converge

[1]   A study of the number of transistors produced annually was undertaken by Gordon Moore – in 2006 this number was placed

at 1018 transistors per annum. Brock, D. (ed.), *Understanding Moore's Law: Four Decades of Innovation,* Chemical Heritage Foundation, 2006.

[2] More information about the Cyber Tyre can be found at: www.us.pirelli.com/web/technology/technology-revolution/cyber-tyre/default.page.

[3] IBM's Smarter Planet: www-07.ibm.com/innovation/au/smarterplanet.

[4] A range of case studies on RFID in supply chains can be found at: www.informationweek.com/news/mobility/RFID/showArticle.jhtml?articleID=163700955.

[5] 'Clever Fridges Co-operate', *Ecogeneration Magazine,* Great Southern Press, May/June 2009.

[6] For information about Planetary Skin, visit: www.planetaryskin.org.

[7] View the original image obtained from the *Explorer VI* Earth satellite at: http://grin.hq.nasa.gov/ABSTRACTS/GPN-2002-000200.html.

[8]   'Digital Mapmakers', *The Economist,* 3 September 2009.

[9]   Aanensen, D., Huntley, D., et al., 'EpiCollect: Linking Smartphones to Web Applications for Epidemiology, Ecology and Community Data Collection', *PLoS ONE,* vol.4, no.9, 2009, available at: www.plosone.org/article/info :doi/10.1371/journal.pone.0006968.

[10]   Information about the NYC Cricket Crawl can be found at: http://pick14.pick.uga.edu/crick et.

[11]   'Sensaris Wearable Sensor Promises to Track Noise and Air Quality', *Engadget Magazine,* 26 May 2009.

[12]   Information about the size of the mobile dating market is given in 'Mobile User-Generated Content: Social Networking, Dating and Personal Content Delivery', Juniper Research, 2009, available at: http://junip erresearch.com. 13. QR code reader software can be downloaded from: scanme.com .au.

[13]   QR code reader software can be downloaded from: scanme.com.au.

[14]   The Layar software can be downloaded from: http://layar.eu.

[15]   An article on the future of digital contact lenses is given in Chen, B., 'Digital Contacts Will Keep an Eye on Your Vital Signs', *Wired Magazine,* 10 September 2009.

[16]   'Wallet of the Future? Your Mobile Phone', CNN.com , 13 August 2009: http://edition.cnn.com/2009/TECH/08/13/cell.phone.wallet.

[17]   'The Power of Mobile Money', *The Economist,* 24 September 2009.

[18]   An example of the resistance to Google Street View was covered in the article, 'Privacy Concerns Delay, Disrupt Google Street View in Europe, Britain', *Los Angeles Times,* 14 September 2009.

[19]   Information about the Tririga Real Estate Environmental

Sustainability product can be found at: www.tririga.com/resources/pdf/TREESFactSheet.pdf.

Chapter 9: Atoms are local, bits are global

[1] A history of the 'Dig for Victory' campaign and the 'Dig for Victory' anthem can be found at: www.bbc.co.uk/dna/h2g2/A2263529.

[2] '1.02 Billion People Hungry', Food and Agricultural Organisation: www.fao.org/news/story/en/item/20568/icode.

[3] For more information on the challenges of food security, see the Global Hunger Index at: www.ifpri.org/publication/2009-global-hunger-index.

[4] Hecht, J., 'First Self-Replicating Machine Makes an Appearance', *New Scientist,* 9 June 2008.

[5] Information about the Objet 3D printer and its applications can be found at: www.objet.com/Applications/Application_Notes.

[6]   An overview of the United States energy grid is given on the US Department of Energy Office of Electricity Delivery and Energy Reliability Gridworks site: http://sites.energetics.com/gridworks/grid.html.

[7]   This and other case studies of long tails can be found on Chris Anderson's blog: www.longtail.com.

[8]   More information about Craigslist can be found at: www.craigslist.org/about/factsheet.

[9]   'Brand Managers' High-wire Act: Going Global and Staying Local', Knowledge@Wharton, Wharton University of Pennsylvania, 14 November 2007.

[10]   An overview of the txteagle project is given at: http://txteagle.com/learn.html.

[11]   Desai, M., 'The Decentering of the Global Firm', Harvard Business School – Finance Unit, National Bureau of Economic Research, 2008.

Chapter 10: If in doubt, look to nature

[1]   'Antarctic Sea Urchin Shows Amazing Energy-Efficiency in Nature's Deep Freeze', *Bio-Medicine,* 7 March 2001.

[2]   Benyus, J., *Biomimicry: Innovation Inspired by Nature,* Harper Collins, 1997. The nine basic laws underpinning biomimicry are:

1) Nature runs on sunlight.

2) Nature uses only the energy it needs.

3) Nature fits form to function.

4) Nature recycles everything.

5) Nature rewards cooperation.

6) Nature banks on diversity.

7) Nature demands local expertise.

8) Nature curbs excesses from within.

9) Nature taps the power of limits.

[3]   Vincent, J., Bogatyreva, O., et al., 'Biomimetics: Its Practice and Theory', *Journal of the Royal*

Society Interface, vol.3, no.9, 22 August 2006, pp471–482.

[4]   Information about the Speedo Fastskin technologies can be found at the Speedo Aqualab site: www.speedo.com/en/aqualab_technologies/aqualab.

[5]   'Shark Skin Saves Naval Industry Money', *Bio-medicine,* 15 July 2005.

[6]   'A Fin-Tuned Design', *The Economist,* 19 November 2008.

[7]   Peeples, L., 'The Secret Right under our (Bottle)Noses: What Dolphins Can Teach us about Hydrodynamics,' *Scientific American,* 29 June 2009.

[8]   The bioSTREAM system is described in detail at: www.biopowersystems.com/biostream.php.

[9]   For more information about the biology of the shark, visit the ReefQuest Centre for Shark Research site at: http://elasmo-research.org/index.html.

[10]   For information about the emerging properties of spider silk, see: 'Better Spider Silk',

The Economist, 29 January 2009; and 'Silk the Key to Rebuilding Bone', *ABC Science,* 18 November 2008.

[11]   Abraham Stroock's work can be found at: www.cheme.cornell.edu/people/profile/index.cfm?netid=ads10.

[12]   The Biomimicry Taxonomy begins at: www.asknature.org/article/view/biomimicry_taxonomy. It also includes a handy one-page taxonomy breakdown.

[13]   Gendall, J., 'Architecture that Imitates Life', *Harvard Magazine,* September–October 2009.

[14]   Frosch, R., and Gallopoulos, N., 'Strategies for Manufacturing', *Scientific American,* vol.261, no.3, 1989, pp144-152.

[15]   Research on the evolution of Wikipedia was undertaken by the Augmented Social Cognition Research Group at the Palo Alto Research Center: http://asc-parc.blogspot.com/2009/08/part-2-more-details-of-changingeditor.html.

[16]   Bonabeau E., and Meyer, C., 'Swarm Intelligence: A Whole New Way to Think About Business', *Harvard Business Review,* May 2001.

[17]   More information about the nutrient load entering the Great Barrier Reef can be found at the Great Barrier Reef Marine Park Authority's website: www.gbrmpa.gov.au/corp_site/key_is sues.

[18]   'The Global Market for Organic Food & Drink: Business Opportunities & Future Outlook', Organic Monitor, November 2006: www.organicmonitor.com /700240.htm.

Epilogue: Eco-natives

[1]   Malthus, T.R., *An Essay on the Principle of Population,* J. Johnson, London, 1798.

[2]   Kremer, M., 'Population Growth and Technological Change: One Million BC to 1990', *Quarterly Journal of Economics,* vol.108, no.3, August 1993, pp681–716.